Genome Editing Tools and Gene Drives

POCKET GUIDES TO
BIOMEDICAL SCIENCES

Series Editor
Lijuan Yuan

A Guide to AIDS, *by Omar Bagasra and Donald Gene Pace*

Tumors and Cancers: Central and Peripheral Nervous Systems, *by Dongyou Liu*

A Guide to Bioethics, *by Emmanuel A. Kornyo*

Tumors and Cancers: Head – Neck – Heart – Lung – Gut, *by Dongyou Liu*

Tumors and Cancers: Skin – Soft Tissue – Bone – Urogenitals, *by Dongyou Liu*

Tumors and Cancers: Endocrine Glands – Blood – Marrow – Lymph, *by Dongyou Liu*

A Guide to Cancer: Origins and Revelations, *by Melford John*

Pocket Guide to Bacterial Infections, *edited by K. Balamurugan*

A Beginner's Guide to Using Open Access Data, *by Saif Aldeen Saleh Airyalat and Shaher Momani*

Pocket Guide to Mycological Diagnosis, *edited by Rossana de Aguiar Cordeiro*

Genome Editing Tools and Gene Drives: A Brief Overview, *by Reagan Mudziwapasi, Ringisai Chekera et al.*

For more information about this series, please visit:
https://www.routledge.com/Pocket-Guides-to-Biomedical-Sciences/book-series/CRCPOCGUITOB

Genome Editing Tools and Gene Drives

A Brief Overview

Reagan Mudziwapasi, Ringisai Chekera, Clophas Zibusiso Ncube, Irvonnie Shoko, Berlinda Ncube, Thandanani Moyo, Jeffrey Godfrey Chimbo, Jemethious Dube, Farai Faustos Mashiri, Moira Amanda Mubani, Duncan Maruta, Charity Chimbo, Mpumuzi Masuku, Ryman Shoko, Rutendo Patricia Nyamusamba, and Fortune Ntengwa Jomane

CRC Press is an imprint of the
Taylor & Francis Group, an **informa** business

First edition published 2022
by CRC Press
6000 Broken Sound Parkway NW, Suite 300, Boca Raton, FL 33487–2742

and by CRC Press
2 Park Square, Milton Park, Abingdon, Oxon, OX14 4RN

CRC Press is an imprint of Taylor & Francis Group, LLC

ISBN: 978-0-367-76066-3 (hbk)
ISBN: 978-0-367-74596-7 (pbk)
ISBN: 978-1-003-16531-6 (ebk)

DOI: 10.1201/9781003165316

Typeset in Frutiger Light
by Apex CoVantage, LLC

I dedicate this book to Tirivashe and Tichatonga Joshua Mudziwapasi, the Mudziwapasi family, my Mandela Washington Fellowship family, Oklahoma State University and all biotechnologists and youth trying to make a positive contribution to society. TOGETHER WE WILL.

Reagan Mudziwapasi

I would like to thank my husband, Dr. Fortune N. Jomane, and my son, Jayden, for their support during the time of writing this book.

Ringisai Chekera

To my husband, Perseverance Shoko, my best friend, greatest support and strongest motivation. Thank you for being there for me.

Irvonie Shoko

I would like to thank God for His blessings and also for making this possible. I also thank my mother (Sithandazile Jamela), grandmother (Julia Sibanda), my siblings (Blessing, Brenda, Iwine, Mbongeni, and Gugulethu), aunt (Filda Gombagomba) and friends (Gari, Themba and Headwick) for their support and encouragement.

Berlinda Ncube

This work is dedicated to my wife (Buhlebenkosi Dube), my daughters (Providence, Thandiwe, and Priviledge) and my son (Prosperity).

Jemethious Dube

This work is dedicated first to God Almighty and secondly to my wife Sakhile and daughters, Unathi and Nyashadzashe for their love and support and also to my parents and siblings for their encouragement.

Farai Faustos Mashiri

Contents

Preface

This book was motivated by the realization that genome editing tools and gene drives can be used to solve many problems bedeviling Africa and the world at large. Genome editing tools and gene drives are revolutionizing science. However, not many African scientists have written on them. In Nelson Mandela's words,

> Education is the great engine of personal development. It is through education that the daughter of a peasant can become a doctor, that the son of a mineworker can become the head of the mine, that a child of farm workers can become the president of a great nation. It is what we make out of what we have, not what we are given, that separates one person from another.

Martin Luther King Jr. also observed that "It is not beyond our power to create a world in which all children have access to a good education. Those who do not believe this have small imaginations". One good act by faculty at Oklahoma State University of giving books to take back to Africa at the end of the Mandela Washington Fellowship motivated the writing of this book. Accordingly, a team of African lecturers and students from various backgrounds and universities in Zimbabwe but with experience in biotechnology was formed to write this book. The goals are to motivate the adoption of genome editing and gene drives and the writing of biotechnology books by African scientists and to provide literature that African students can identify with, that is easy to follow and that is usable in teaching. Additionally, the book is an attempt to allay some fears associated with genome editing and gene drives by creating a more informed audience and shedding light on the processes. Consequently, the writing of the book commenced toward the last quarter of 2019 right through 2021. This book talks about the genome editing tools, megaTALs, TALENs, ZFNs and CRISPR. It discusses the principles behind them, their applications and concerns about their use. It also does the same for various gene drives and gene drive systems. The book includes examples of the use of genome editing tools and gene drives in solving societal problems, including those examples which can be beneficial in African settings. Provisions under the Creative Commons greatly aided the completion of this book. This is because figures available under the CC BY license were the only ones we could use in the absence of a budget for the book. Many initially selected images had to be removed from the book due to copyright issues. An increasing number of patents based on the use of genome editing and gene drives are likely to come out in the near future. This book will help scientists understand genome editing and gene drives, make informed decisions and participate in the patenting race, realizing economic gains for self and country while helping humanity. Currently, the patenting race is skewed in favor of developed countries, although developing countries such as many in Africa can easily take the lead due to their large biodiversity. Through difficult deadlines and obstacles, the authors remained committed, and now their contribution to society through this book is out. "No country can really develop unless its citizens are educated," Mandela said. This is part of our contribution to the development of Africa and the world based on the use of genome editing and gene drives.

Acknowledgments

We would like to acknowledge the eye-opening experience provided through the Mandela Washington Fellowship for Young African Leaders to the lead author. This experience greatly influenced the writing of this book. We would also like to acknowledge Oklahoma State University, the institute through which the training was provided. In particular, we would like to acknowledge Professor Craig Edwards and his team, the Department of Agricultural Education, Communications and Leadership, Professor Craig Watters and his team and the Spears Business School. Those many books you gave us to share with other Africans back home motivated the writing of this book. We would also like to acknowledge the help given by Dr. Fanuel Songwe and Mr. Vitaris Kodogo during the preparation of this book. The opportunities and access to infrastructure afforded by Lupane State University are greatly appreciated.

Authors

Ringisai Chekera is an animal scientist who specializes in animal breeding. She holds a BSc in agriculture (animal science) and an MSc in agriculture (animal breeding and biotechnology). She has a great passion for building the resilience of livestock farmers through the use of various breeding strategies. She has contributed to the Resilience Knowledge Hub as a researcher in a project implemented under the ZRBF project to enhance the resilience of smallholder poultry farmers.

Charity Chimbo has been a biology and agriculture science educator since 2010. She holds a BSc honors degree in agriculture (University of Zimbabwe) and a postgraduate diploma in education (Zimbabwe Open University) and is studying toward an MSc in animal breeding and biotechnology with Lupane State University. Her research interests are in genomics and bioinformatics.

Jemethious Dube is an agricultural scientist. He holds a diploma in agriculture from Esigodini College, a BSc in agriculture management, a BSc special honors degree in crop science and a master's degree in business administration, all from the Zimbabwe Open University. He also holds a postgraduate diploma in education from Lupane State University. He is studying toward an MSc in crop science (plant breeding) with Lupane State University. He has vast experience working with financial institutions and nongovernmental organizations. He is passionate about fieldwork.

Fortune Ntengwa Jomane is an animal scientist who specializes in animal breeding and genetics. He holds a BSc in agriculture (animal science), an MSc in agriculture (animal science) and a PhD in agriculture specializing in animal breeding and genetics. His passion is enhancing animal production through exploiting genetics. He has published several peer-reviewed papers in animal breeding and genetics. He is a senior lecturer and chairperson of the Department of Animal Science and Rangeland Management at Lupane State University. He is the secretary for the Beef and Leather Value Chain Technical Assistance Project implemented by the government of Zimbabwe through the Ministry of Industry and Commerce and the Ministry of Lands, Agriculture, Fisheries, Water and Rural Resettlement.

Farai Faustos Mashiri has been an agricultural science and biology educator since 2009 at various institutions in Zimbabwe and recently South Africa. He received his BSc in agriculture and natural resources in 1999 from Africa University and post graduate diploma in education from Lupane State University. After graduating, his early career engagements included technical work for private enterprises in the agricultural sector, mainly in aquaculture and horticulture. He is currently undertaking postgraduate studies in animal breeding and biotechnology. His research interests are in tilapia breeding, genome editing and gene drives. His areas of interest also span the fields of bioinformatics, genomics and data analysis using R and Python.

Mpumuzi Masuku is a final year student at Lupane State University studying towards a master of science degree in animal breeding and biotechnology. He holds a BSc honors degree in agriculture (animal science) from the University of Zimbabwe. He is detail-oriented and values self-actualization immensely. He aims to leverage proven knowledge of research in animal breeding, molecular biotechnology and genomics.

Thandanani Moyo obtained a BSc (Hons) in applied biology and biochemistry from NUST, Zimbabwe. Currently, he is studying toward an MSc in animal breeding and biotechnology at Lupane State University. His research interests are in molecular breeding and animal health. He is currently working as a lab scientist at a diagnostics laboratory.

Moira Amanda Mubani is a talented and hardworking young scientist. She holds a BSc honors degree in biotechnology and is currently in her final year completing her master's degree in applied pharmaceutical science. Her interests incude computational chemistry, where she developed ways of analyzing a substantial amount of data. She uses molecular modeling and bioinformatics to design novel anti-tuberculosis and anticancer drugs. She also works on ethnopharmaceuticals projects, where she is involved in cannabis research. She has been working at the National Biotechnology Authority (NBA) since October 2019. She was seconded by NBA to work at the African Institute of Biomedical Science and Technology (AiBST) in 2020 and was recently seconded to work with Dr. Justen Manasa at the University of Zimbabwe Innovation Hub. Mubani was among the 10 people selected to participate in a hands-on workshop training in advanced genomics technologies organized by the Biomedical Research and Training Institute.

Reagan Mudziwapasi is a biotechnology researcher, lecturer and entrepreneur. He is a cofounder and Operations Director at EnviroBiotech Solutions Africa, a company that develops biotechnology-based solutions. He has authored several peer-reviewed journals, is a patent holder, and has several inventions and has conducted several consultancies. He has contributed to various discussions on biotechnology, entrepreneurship and energy *inter alia*. He promotes awareness on genetic engineering and its safe and responsible use. He helps marginalized communities and upcoming inventors protect their intellectual property.

Berlinda Ncube is a cofounder of Go Science Club where she helps learners develop science projects. She joined Zezani High School in 2018 as a chemistry teacher. She is a graduate of the National University of Science and Technology where she obtained a BSc Hons in applied biology and biochemistry and Zimbabwe Open University where she obtained a postgraduate diploma in education. She is currently working towards obtaining an MSc in applied animal breeding and biotechnology at Lupane State University.

Clophas Zibusiso Ncube is an animal science and agribusiness practitioner. He has over ten years of continuous work experience. In 2018, he was awarded the Australian Awards Scholarship to study agribusiness concentrating on value chains. He holds a master of science degree in agribusiness from Africa University in Mutare, Zimbabwe, a BSc in agricultural science and a BSc in agricultural science honors degree in animal science. He is currently a student for the master of science degree in animal breeding and biotechnology at Lupane State University.

Rutendo Patricia Nyamusamba has a passion for development and being a voice for the seemingly voiceless. It is this passion that drives her desire for research. She uses a holistic approach in solving food system challenges. Her research and professional background are strongly rooted in agronomy and physiology of field crops, focusing on sustainable crop production and weed science. She is currently working on the integration of livestock in cropping systems for both commercial and small-scale farmers. She has a PhD in agronomy with a minor in statistics; an MSc in plant science from South Dakota State University, USA, and a BSc in agriculture with an agribusiness specialty from Africa University in Mutare, Zimbabwe. She teaches modules in weed science, plant physiology, crop production, sustainable agriculture and statistics. Occasionally, she assists with some specific modules in agricultural economics such as econometrics.

Irvonnie Shoko is a founder of Divin Chen Agribusiness Solutions, where she offers consultancy and training services in livestock production. She joined Midlands State University in 2019 as a part-time lecturer in animal breeding and physiology. Prior to that, she was a research scientist at SIRDC. During her tenure at SIRDC, she helped the company to set up a robust livestock breeding and production project. She is a graduate of the University of Zimbabwe, where she obtained a BSc in animal science and an MSc in biotechnology. She is currently working toward an MSc in animal breeding and biotechnology at Lupane State University.

Ryman Shoko is a senior lecturer in the Department of Biological Sciences at the Chinhoyi University of Technology. His areas of expertise include molecular biology, proteomics bioinformatics and computational systems biology.

Abbreviations

AAV Adeno-Associated Viruses
ABNE African Biosafety Network of Expertise
AIDS Acquired Immunodeficiency Syndrome
AMRH African Medicines Regulatory Harmonization
ATP Adinosine Triphosphate
ATTR Amyloid Transthyretin
CAR Chimeric Antigen Receptor
CF Cystic Fibrosis
CFTR Cystic Fibrosis Transmembrane Conductor Receptor
CHO Chinese Hamster Ovary
CI Cytoplasmic Incompatibility
CRISPR Clustered Regularly Interspaced Short Palindromic Repeats
DGAT1 Diacylglycerol Acyltransferase-1
DHFr Dihydrofolate Reductase
DMD Duchenne Muscular Dystrophy
DNA Deoxyribonucleic Acid
DRs Direct Repeats
DSB Double-Strand Breaks
ENMOD Environmental Modification
ES Embryonic Stem
GDOs Gene Drive Organisms
GEs Genome Editors
GMO Genetically Modified Organism
HAs Homology Arms
HDR Homology-Directed Repair
HEGs Homing Endonuclease Genes
HIV Human Immunodeficiency Virus
HR Homologous Recombination
HSC Hematopoietic Stem Cells
HSPC Hematopoietic Stem and Progenitor Cells
HTGTS High Throughput Genome-wide Translocation Sequencing
HUGO Human Genome Organization
IIP Incompatible Insect Technique
INDELs Insertions and/or Deletions
IPSCs Induced Pluripotent Stem Cells
LCTR Large Clusters of Tandem Repeat
LTRs Long Terminal Repeats
MCR Mutagenic Chain Reaction
MEDEA Maternal-Effect Dominant Embryonic Arrest
MK Male Killing
MMEJ Microhomology-Mediated End Joining
NEPAD New Partnership for Africa's Development
NHEJ Nonhomologous End Joining
NLS Nuclear Localization Signal
PAM Protospacer Adjacent Motif
PAR Poly-ADP-Ribose
PBSC Peripheral Blood Stem Cells
PCR Polymerase Chain Reaction
PHD Plant Homeodomain
PI Parthenogenesis Induction
ReMOT Receptor-Mediated Ovary Transduction
RGENs RNA-Guided Engineered Nucleases
RNA Ribonucleic Acid
RNP Ribonucleoprotein
RVDs Repeat Variable Diresidues
SCID Severe Combined Immune Deficiency
SELEX Systematic Evolution of Ligands by Exponential Enrichment

SIT Sterile Insect Technique
TAL Transcription Activator-Like
TALENs Transcription Activator-Like Effector Nucleases
TALEs Transcription Activator-Like Effectors
TEs Transposable Elements
TGR Target Gene Replacement
TM Targeted Mutagenesis
UN United Nations
WMP World Mosquito Program
WNV West Nile Virus
WTO World Trade Organization
ZF Zinc Finger
ZFNs Zinc Finger Nucleases
ZFPs Zinc Finger Proteins

1

Genome Editing

History

The interaction of the environment and the genotype of an organism determine its characteristics. A complete set of an organism's genetic material is called its genome. The genome is composed of RNA and DNA (Ahmed et al., 2019). Microbes, animals, and plants also typically have either RNA or DNA or both as their genome. All the genetic information for the growth and development of the organism is contained in the genome (The Royal Society, 2016). A long time before the field of genetics was established, humans genetically modified plants through breeding and selection. Without knowledge of genes, mutagenesis, or gene editing, people of old influenced the genetic makeup of plants and animals by selecting for traits that are conducive to food production and culling those that had unfavorable traits (Wang et al., 1999).

Genome editing tools are tools that can be used to edit the genome of any organism by manipulation of the specific gene loci to gain genome modifications, such as insertions, deletions, or point mutations (Baker, 2012). Conventional methods that have been used to alter genomes or genetic material have been largely based on the use of chemicals and radiation. However, imprecise changes to the genetic material are made using these methods, making it difficult to predict results from the modification events. The advent of recombinant DNA technology in the 1970s gave scientists the ability to make more precise changes to genetic material by inserting or deleting genes or bases. Nevertheless, the process was still relatively imprecise and consequently made inaccurate changes to the genetic material.

In the olden days, genetic studies relied on the discovery and analysis of spontaneous mutations. In the mid-twentieth century, Muller (Muller, 1927) and Auerbach (Auerbach et al., 1947) demonstrated that the rate of mutagenesis could be enhanced with radiation or chemical treatment. Later on, methods that relied on transposon insertions that could be induced in some organisms were used, but these procedures, like radiation and chemical mutagenesis, for example, produced changes at random sites in the genomes of organisms. The first targeted genomic changes were produced in yeast and mice in the 1970s and 1980s. This gene targeting depended on the process of homologous recombination, which was remarkably precise but very inefficient in its outcome, for example, in mouse cells. Recovery of the desired products required a powerful selection and thorough characterization. Because of the low frequency and the absence of culturable embryonic stem cells in mammals other than mice, this resulted in gene targeting not being readily adaptable to other species (Thomas et al., 1986). Traditionally, a poor understanding of genomes contributed to the ineffectiveness of gene editing using these tools. The situation, however, changed with the improvement of technology and subsequent sequencing of genomes of various organisms. It is noteworthy that techniques for gene editing that can make more precise and intentional changes to genes and genomes were later developed and are in use (Robb, 2019; Qi, 2017). These techniques can accurately edit, remove, or insert bases or genes in genetic material or genomes (Urnov, 2018). All the gene-editing technologies utilize DNA-cutting enzymes called nucleases. They also have a DNA-targeting mechanism that guides the nuclease to a specific location on the DNA that has to be cut (Zhang et al., 2018). These mechanisms usually involve proteins that make it easy for gene-editing tools to manipulate specific genes in certain areas, as opposed to random gene changes that were made by prior tools (Malzahn et al., 2017). These gene-editing tools are capable of modifying one gene at a time. They typically modify (edit), delete, or insert bases. More advanced gene-editing tools such as those based on the clustered regularly interspaced short palindromic repeats (CRISPR) are capable of editing multiple genes simultaneously. Editing of the genome is referred to as genome editing (Winterberg et al., 2019). These tools were developed in the last 10 years and are still being perfected, although they have high gene-editing efficiencies. Their efficiency is also attributed to the fact that genome sequences for most species are now easily assembled and available. These are used in targeting and guiding the endonucleases (Koch, 2016). Zinc

DOI: 10.1201/9781003165316-1

finger nucleases (ZFNs), megaTALs, and transcription activator-like effector nucleases (TALENs) recognize protein and DNA, while CRISPR recognizes RNA and DNA. These facilitate the editing of genes and altering pathways giving scientists the ability to micro-edit DNA codes and mRNA fate via posttranscriptional modifications (Wright et al., 2018). Currently, we have three powerful classes of nucleases that can be programmed to make double-strand breaks (DSBs) at essentially any desired target: ZFNs, TALENs, and CRISPR-Cas. However, CRISPR-Cas now dominates in research laboratories around the world; the other two are also still in use for research and various agricultural and medical research. All of these nucleases arose from investigations into natural biological processes and not from intentions to find genome editing reagents (Carroll, 2014). CRISPR-Cas9 is the most common technology being applied now because it has become cheaper, faster, more accurate, and more efficient than other existing genome editing methods. The Cas9 endonuclease is a system made up of four parts that include two small RNA molecules named CRISPR RNA (crRNA) and trans-activating CRISPR RNA (tracrRNA) (Barrangou, 2014).

Principle

The foundation of the development of targeted genome editing is based on DNA repair mechanisms after its damage and the consequent structural changes in the DNA. A combination of regulatory proteins and sequence-identifiable programmable nucleases makes site-specific genetic and epigenetic regulations possible (Guha et al., 2017). In the last decade, a blooming of targeted genome editing tools and applications have been witnessed in the research community. Programmable DNA nucleases such as ZFNs, TALENs, and the clustered regularly interspaced short palindromic repeats-Cas9 system (CRISPR-Cas9) possess long recognition sites and they can cut DNA in a very specific manner. These DNA nucleases facilitate targeted genetic alterations by enhancing the DNA mutation rate via the introduction of double-strand breaks (DSB) (Source: Roy et al., 2018) at a predetermined genomic site. Figure 1.1 shows DNA editing using tools such as CRISPR, ZFN, and TALENs.

A designer nuclease (ZFN, TALEN, or CRISPR-Cas9) cleaves a DNA molecule at a target DNA site to generate a DSB. The DSB can be repaired with one of two endogenous DNA repair mechanisms. These are the nonhomologous end joining (NHEJ) and the homology-directed repair (HDR). The two ends of the DSB are joined together in the NHEJ pathway and ligated without a homologous repair template, which often inserts or deletes nucleotides (indels) to cause gene disruption (knockout). The HDR pathway requires an exogenous DNA template to be provided along with a site-specific genome editing nuclease to repair the DSB, thereby

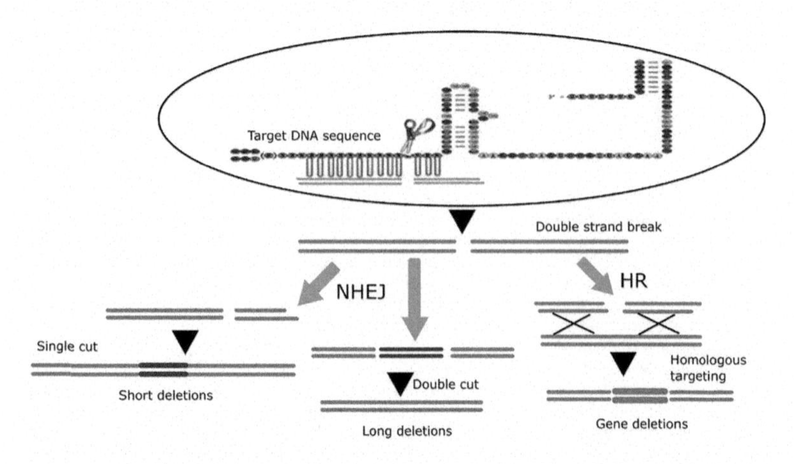

Figure 1.1 Gene (genome) editing of DNA using tools such as ZFN, TALEN, or CRISPR-Cas9 techniques.

triggering the knock-in of the desired DNA sequence into the genome of an embryo or animal cells (Chandrasegaran and Carroll, 2016).

Compared to conventional homologous recombination-based gene targeting, DNA nucleases, which are also referred to as genome editors (GEs), can increase the targeting rate by around 10,000- to 100,000-fold. The successful application of different genome editors has been shown in a range of different organisms, for example in insects, amphibians, plants, nematodes, and several mammalian species, including human cells and embryos (Gurdon and Melton, 2008). Gene-editing technologies that use programmable nucleases include ZFNs, TALENs, and CRISPR. The CRISPR nucleases are also called RNA-guided engineered nucleases (RGENs). The differences among these are that ZFNs are hybrids between a DNA cleavage domain from a bacterial protein and sets of zinc fingers that were originally identified in sequence-specific eukaryotic transcription factors. TALENs employ the same bacterial cleavage domain but link it to DNA recognition modules from transcription factors produced by plant pathogenic bacteria. CRISPR-Cas is a prokaryotic system of acquired immunity to invading DNA or RNA. ZFNs were a result of the first eukaryotic sequence-specific transcription factor to be characterized, which was found to have zinc-binding repeats in its DNA-binding domain. Related sequences from other transcription factors were shown to be peptide modules (Pavletich and Pabo, 1991). Changing a few residues in a single zinc finger altered its DNA recognition specificity, and fingers could be devised to recognize many different DNA triplets. In the case of TALENs, some plant pathogenic bacteria secrete into host cells proteins that bind to and regulate the activity of host genes to promote the infection. For ZFNs and TALENs, some bacterial restriction enzymes cut DNA a few base pairs away from their recognition sites, and this is because they have physically separable binding and cleavage domains. CRISPR-Cas began with the discovery of a cluster of odd, short repeats in a bacterial genome. Between those CRISPR sequences are short sequences that were shown to match viral genomes (Barrangou et al., 2007). Some CRISPR-associated (Cas) proteins encoded adjacent to the repeat clusters mediate capture of these viral sequences, while others mediate cleavage and inactivation of invading viral genomes, guided by short RNAs (crRNAs) transcribed from the CRISPR arrays. The identification of the small trans-acting RNA (tracrRNA) that participates in both processing of the crRNAs and cleavage of the invading DNA in *Streptococcus pyogenes* was a great discovery in the process. Putting together the crRNA with tracrRNA and the one protein needed for cleavage in this system (Cas9) led to the editing reagent that is now most widely used in most researches (Barrangou and Marraffini, 2014). These tools make it possible to perform genome editing in whole organisms and cultured cells of any kingdom (Zhang et al., 2019). The cell repair machinery of the cells whose genome would have been cut is responsible for its repair. However, if instructions for repair are not provided for the cell repair machinery, mutations can occur. When instructions for repair are available, they will be followed during the genome repair. Researchers manipulate the cell repair machinery by providing different repair information from what the cell would affect (Wright et al., 2018). This makes it possible to modify genomes, for example, by replacing mutated forms of genes that cause Parkinson's disease or genetic metabolic disorders with those that do not (Soldner, 2016; Hao, 2014). Genes conferring better disease resistance to, for example, powdery mildew can also be used to replace those that proffer less resistance (Gao, 2018). Genome editing techniques can modify genes without the introduction of foreign DNA. This is the case in some approaches used for switching off genes of lab-grown cells to determine their function. This can also be done by switching off disease-causing genes (Maeder and Gersbach, 2016).

Applications of Genome Editing

Currently, the world over, almost 1 billion people suffer from chronic malnourishment, while at the same time our agricultural systems are under pressure and degrading, facilitated by the loss of biodiversity and the increasing uncertainties of climate change (Foley et al., 2011). With the global population projected to exceed 9 billion by 2050, conventional agriculture will face huge challenges, thus requiring crops with higher yields and of improved nutritional quality and at the same time requiring fewer inputs to reduce costs of production (Tilman et al., 2011). Although conventional breeding is currently the most widely used approach in crop improvement, it is labor-intensive, and it takes a very long time, for example, several years to progress from the early stages of screening phenotypes and genotypes to the first crosses to come up with commercial varieties. Many benefits can accrue to economies as a result of adopting genome editing technologies. These benefits are mainly in agriculture, health, and environmental protection, as indicated in Table 1.1. Genome editing can increase food security, agricultural productivity, and the nutritional quality of foods (Mir et al., 2018). Additionally, it can also make species more resistant to disease and climate variations. Genome editing can be manipulated to improve industrial bioprocesses and to harness biofuels (Shukla-Jones et al., 2018). Rapid growth is being experienced in the genome editing field. This is partly due to more new methods and technologies for genome editing that are emerging.

The improvements in genome editing tools enhanced the accelerated nuclease assembly and nuclease performance. These methods are now more efficient, easier, and cheaper. Consequently, new approaches to increasing areas of biotechnology such as biopharmaceutical production, preclinical and clinical gene therapy for disease treatment, transgenic organisms and cell lines, genome structure, regulation, and function are being advanced (Yao, 2018). CRISPR is largely accepted due to its high modification efficiency (Gao, 2018). It has unsurpassed insertion and deletion (indel) efficiency in a variety of cells compared to other programmable nucleases (Moon et al., 2019).

Boosting agricultural productivity and food safety is of immense importance because the world population is expected to grow to about 9.6 billion by the year 2050, as alluded to earlier. Concurrently, the amount of arable land is decreasing (Malzahn et al., 2017; Martínez-Fortún et al., 2017). An example of employing genome editing tools toward this end is using the CRISPR-Cas technique, the genes encoding for ovalbumin and ovomucoid have been knocked out to remove the two major allergenic components from the egg white. This could make eggs digestible for a wider range of consumers that could otherwise not consume chicken eggs because of allergic reactions (Oishi et al., 2016).

Genome editing techniques, for example, ZFNs, TALENs, and RNA-guided DNA endonucleases (CRISPR-Cas), have been shown to revolutionize biological research with great benefits for personalized medicine. These emerging technologies significantly expand the ability to come up with and study model organisms, including large animals, and they will play a crucial role in correcting many genetic diseases in livestock species and humans. With the availability of these tools, researchers can develop biomedical models in species that are more physiologically closely related to humans than mice are. The domestic pig is particularly promising in this regard. The growing number of human disease models in pigs supports this assumption (Flisikowska et al., 2014). Due to the physiological similarity with humans, which is high, porcine organs are considered as critical potential solutions to meet the growing demand of human organs for allotransplantation. To achieve the goal of allotransplantation and to avoid immune rejection responses in humans, the porcine genome has to be modified to ensure the long-term survival of the porcine organs in patients after xenografting. ZFNs, TALENs, and CRISPR-Cas can now be used to efficiently knockout candidate pig genes or to precisely knock-in transgenes at specific genomic sites in the porcine genome to produce pigs specifically custom made as organ donors (Flisikowska et al., 2014). Key benefits and applications of gene-editing technologies are summarized in Table 1.1.

Table 1.1 Key Benefits and Applications of Gene-Editing Technologies

Agriculture	Environmental Conservation	Energy Production	Health Care
Food security	Endangered and extinct species	Improved economic output	Improved human health
	Biodiversity	Clean energy	Enhancement
			Reduced expenditures on health care and disability
Mechanisms	Mechanisms	Mechanisms	Mechanisms
Increased production	Disease- and climate-resistant species	Industrial bioprocesses	Diagnostics
Climate-smart crops and animals	Invasive species control	Biofuels	Disease modeling Treating genetic diseases Tissue transplants Regenerative medicine

Source: Shukla-Jones et al., 2018.

Table 1.2 Examples of Successful Implementation of Genome Editing for Crop Improvement

Crop	Gene	Function	Methods	Tools
Corn	ALS1, ALs2	Herbicide resistance	Promoter disruption	CRISPR-Cas9
Cotton	hppd, epsps	Herbicide resistance	Promoter disruption	Meganuclease
Wheat	TaMLO	Disease resistance	Gene knockout	TALEN

Many different applications are being pursued using gene-editing techniques and the only limit will be our imagination. However, there is a need to consider what are the best uses of the technology, what adjustments are needed to make the technology safe and effective in different sectors, and how its advances will be provided to those who would benefit most from the technology (Kaiser, 2016). Some examples of where genome editing was successfully used in crop improvement are listed in Table 1.2.

The encouraging results and applications of CRISPR tools in agriculture have already been shown in crops such as wheat, maize, and tomatoes. SDN-1 in wheat is used to deal with devastating powdery mildew fungi, while more challenging, complex traits have been changed in corn and tomato. In maize, the application of SDN-3 to the Argos8 (also known as Zar8) gene promoter presented a constitutive expression of the endogenous gene resulting in increased yields of maize during water shortage stress (Mao et al., 2019). SDN-1 was used to generate mutations in the regulatory regions of the tomato yield genes, which increased their genetic variation and increased yield in a short period of time (Rodríguez-Leal et al., 2017). CRISPR has opened the potential to transform the world around us to support human health and the general environment. Applications include modification or even eradication of insect distributions, such as mosquitoes.

In an endeavor to address economic and animal welfare issues, dairy cows have been bred that lack horns, due to genetic modification. Cows, sheep, pigs, and other food animals that carry more muscle mass (double muscles) have also been produced by disruption of a single gene. Genome editing has the advantage over natural breeding and selection; for example, a trait can be introduced in a single generation without disrupting a favorable genetic background. The same beneficial modification can be introduced into different breeds of animals that are adapted to different environments without leading to monoculture (Proudfoot et al., 2015).

Apart from food modifications, large animal models for human diseases are being produced to facilitate physiological analysis, drug testing, and other therapies for different ailments. Disease conditions that can be addressed using NHEJ-mediated genome editing are those in which mutating a genetic element, whether a coding region, a regulatory element or some other genetic element, might result in a clinical benefit to humans (Li et al., 2020). One example of this approach is to delete the erythroid enhancer for Bcl11A in hematopoietic stem/progenitor cells (HSPCs) to upregulate γ-globin to treat sickle cell disease and β-thalassemia. Both sickle cell disease and β-thalassemia are monogenic diseases caused by mutations in the HBB gene (Bauer et al., 2013).

There are more than 10,000 diseases created from mutations in a single gene (monogenic diseases), and there are even more polygenic diseases or diseases caused by mutations in multiple genes. Because of these diseases and so many deaths as a result, there is a huge need to rectify these most prevalent diseases. With the use and help of CRISPR gene-editing technology, which can turn off genes responsible for some of these diseases, the human race could have a nearly disease-free future. Currently, most of the research involving genetic editing and fixing genetic diseases is on monogenic ones, and of these many monogenic diseases, 14 are being researched to find a solution through gene editing and manipulation. Examples of some of these diseases include amyloid transthyretin (ATTR) amyloidosis, beta-thalassemia, hemophilia type B, and Leber congenital amaurosis 10 (LCA10) (Hsu et al., 2014).

Before the emergence of gene-editing engineered nucleases, genetically modifying mammalian cell lines was labor-intensive, costly, and limited to laboratories with specialized expertise and equipment. However, with the coming up of cost-effective and user-friendly gene-editing technologies, custom cell lines carrying nearly any genomic modification can now be generated in a short space of time. For example, some of the outcomes that have become routine because of the emergence of targeted nucleases include gene deletion (Lee et al., 2010).

It seems likely that genome editing will be applied to companion species, generating new breeds of dogs and cats and correcting genetic susceptibilities in current breeds of companion animals. Additional work will be needed to uncover the genetic causes for desirable traits; currently, genetic research in dogs is at an advanced level and making good progress (Gantz and Bier, 2016).

Concerns with Genome Editing

The editing of DNA means the irreversible permanent change of genome information, and this process is also facing inevitable security risks and ethical problems. All of the nucleases used in genome editing can be very effective, but none of them has perfect specificity. Recent modifications of Cas9 protein and guide RNA (gRNA) have enhanced their discrimination against off-targets. The major concern about off-target cleavage and mutagenesis depends greatly on the application. In many model organisms, there are ways to investigate the effects of an introduced sequence change, for example, making independent mutations in the same gene, crossing into a clean background, and complementing with a wild-type gene (Ahmed et al., 2019). In cases of organisms that can be rapidly expanded, for example, crop plants, founder genomes can be sequenced, and founder phenotypes can be analyzed thoroughly. Even in some medical applications, off-target mutations may be tolerated as long as they do not lead to novel clinical conditions (Tycko et al., 2016). There is generally a low acceptance of genetically modified organisms (GMOs) globally. However, this is directed at mostly foodstuffs and less to other products. Consumers are afraid that transgenes in GMOs can have adverse effects on their health and the environment. However, some genome editing tools achieve transgene-free genome modifications efficiently (Malzahn et al., 2017). Despite these successes by ZFNs, TALENs, and CRISPR-Cas9, there are still challenges concerning their effectiveness, specificity, and safety (Guo et al., 2018). The optimal RNA scaffold of CRISPR-Cas9 for several eukaryotic systems is unknown (Mir et al., 2018).

Some genetic tools such as CRISPR incorporate prokaryote-derived biomolecules into eukaryotic bodies and cells. This raises concerns about the potential of foreign substances to elicit immune responses and cause cellular toxicity. Research shows eukaryotic organisms can produce antibodies against the CRISPR components when they are used to edit their genomes. Blood samples of 79% of participants who received CRISPR assemblies for genome editing had antibodies against SaCas9. Additionally, about 65% were positive for antibodies against SpCas9 (Charlesworth, 2019).

Off-target mutations refer to genome editing at unintended sites. They can emanate from misguides by gRNA or gRNA-independent manners (Zuo, 2019). The potential to have off-target mutations, especially in therapeutic applications of genome editing tools, is still a major concern for many people. There are attempts to improve tools to reduce off-target mutations. One approach is to perform whole-genome sequencing after a genetic modification event to detect the off-target mutations using bioinformatics tools such as sequence alignment.

Despite the potential applications of genome editing in the medical field, guidelines for designing preclinical trial studies still need to be addressed in many countries. The ethics concerning population-level genomic analysis and genome editing are also largely unaddressed (Ma et al., 2017).

There are ethical concerns when it comes to genetically editing human genomes, despite the potential for curing most of the genetic diseases that plague a lot of people. One of the biggest concerns is that it will lead to designer humans whose genes have been edited to give them superior intelligence and other favorable traits. This type of engineering would likely be expensive. Others are opposed to CRISPR gene editing because it is permanent, and those snipped segments of DNA cannot be inverted and would be distributed down to future generations, and there would not be a reversal of the effects. Through these gene-editing technologies, there is fear that there is always room for mistakes and mutations that would need to be contained or corrected. As with other medical advances, each such technology comes with its own set of benefits, risks, regulatory frameworks, ethical issues, and societal implications. Important questions raised concerning genome editing include how to balance potential benefits against the risk of unintended harm to humans; how to govern the use of these technologies; how to incorporate societal values into salient clinical and policy considerations; and how to respect the inevitable differences, rooted in national cultures, that will shape perspectives on whether and how to use these technologies effectively and efficiently (Puchta, 2005).

There are practical, ethical, and legal considerations that need to be fully addressed before genomic-editing technologies are integrated into different areas, for example, in active conservation practices (WTO, 2018). Researchers, practitioners, and policymakers must work together and identify the best and safe approaches for utilizing this technology while also acknowledging that great care must be taken to avoid irreversible harm to organisms at large (Au, 2015).

2

Zinc Finger Nucleases

History of Zinc Finger Nucleases

The capability to modify the complex human genome precisely and efficiently has been the greatest breakthrough that transformed biology and revolutionized the gene therapy arena (Cathomen and Joung, 2008). The main fundamental technology is based on the cellular homologous recombination (HR) pathway. According to Cathomen and Joung (2008), this method is known as gene targeting. This pathway was invented mainly to stimulate genetic recombination in the course of meiosis. It is also expedient in the repair process of DNA double-strand breaks (DSBs) before mitosis.

ZFNs are described by Rémy et al. (2010) as hybrid molecules composed of a designed polymeric zinc finger domain specific for a DNA target sequence and a FokI nuclease cleavage domain. FokI has a requirement of the dimerization to expurgate DNA. The binding of two heterodimers of designed ZFN-FokI hybrid molecules to two adjoining target sequences in each DNA strand disjointed by a 6 base-pair cleavage site results in FokI dimerization and ensuing DNA cleavage (Rémy et al., 2010). ZFNs are the first generation of artificial restriction enzymes that were used in genome editing (Carroll, 2011; Greiner et al., 2017; Ran et al., 2018). In the late 1980s, the first ZFNs were identified (Cassandri et al., 2017). They have between three and six zinc finger DNA-binding domains. These are fused to the heterodimeric nuclease domain FokI, which recognizes five bp of DNA (Greiner et al., 2017; Ran et al., 2018). These ZFNs can be designed to introduce a DSB into a preferred target locus and, as a result, encourage gene targeting 100- to 10,000-fold by initiating cellular DNA repair pathways (Cathomen and Joung, 2008). Transcription factor IIIa (TFIIIa) from *Xenopus laevis* was the first ZFN. ZFNs are among the most abundant groups of proteins. They can interact with RNA, DNA, PAR (poly-ADP-ribose), and other proteins. Thus they have a very wide range of molecular functions (Cassandri et al., 2017).

Urnov et al. (2010) asserts that to be beneficial for genome engineering, an endonuclease must have precise recognition of extended target sequences combined with enough adaptability for retargeting to user-defined sequences. The ZFN architecture is said to meet these specifications by joining the DNA-binding domain of a versatile class of eukaryotic transcription features called zinc finger proteins (ZFPs) with the nuclease domain of the FokI restriction enzyme. The ZFN's structure enables them to combine the favorable merits of both components of the DNA binding, specificity and the flexibility of ZFPs along with a cleavage activity that is vigorous but restrained in the lack of a specific binding event, while retaining functional modularity. As a consequence, both the DNA-binding and catalytic domains can be augmented in isolation, which simplifies retargeting and platform enhancement efforts.

It was stated earlier that ZFNs are hybrid molecules composed of a polymeric zinc finger domain and a FokI nuclease cleavage domain. These domains determine the specificity of ZFNs. Two major platforms exist for generating polymeric zinc fingers with defined specificities. These are the patented platform developed by Sangamo Biosciences and the OPEN platform developed by the Zinc Finger Consortium (Rémy et al., 2010). Both are now available for transgenic animal investigation purposes. Sangamo has partnered with Sigma to sell preassembled ZFNs by using the Compozr program (www.compozrzfn.com/), whereas the OPEN platform (www.addgene.org/zfc; www.zincfingers.org/software-tools.htm) makes their modular assembly zinc finger pools and reagents liberally accessible.

Generation of a DNA DSB by a ZFN results in the instigation of a cellular response well-known as the DNA damage response (Rémy et al., 2010). It is well documented that a DSB can be

DOI: 10.1201/9781003165316-2

fixed in two different ways. One is NHEJ, which generates short insertions or deletions at the cleavage site. It also can be repaired by homologous recombination (HR) using a DNA template, which results in gene knock-ins that are either a faultless repair or, if a modified template is introduced, sequence replacement. According to Rémy et al. (2010), ZFNs have been put to use to generate gene-targeted knock-in in cultured cells in human embryonic stem (ES) cells and induced pluripotent stem (iPS) cells and *Drosophila* as well as gene-targeted knockouts in *Xenopus*, *C. elegans*, and zebrafish. Zinc fingers fused to transcription activators have also been used to target specific promoter sequences and prompt gene expression in transgenic mice. ZFNs have also been used to induce gene-targeted knock-ins in cells in the context of genome editing for gene therapy. Related technologies such as homing endonucleases utilize a different DNA cleavage mechanism, but the resulting DNA DSB is resolved over and done with similar mechanisms, and engineered homing endonucleases or synthetic nucleases can be applied in the same manner as ZFNs. The higher engineering obstacles to generating novel homing endonucleases have so far restricted their use to a few plant and human applications, even though recent progress ought to permit additional extensive application as the engineering hurdles to developing novel homing endonucleases are reduced (Rémy et al., 2010).

Target Cells (Prokaryotic Cells and Eukaryotic Cells)

The ZFN technique has been used to disrupt native loci in model organisms such as rats and *Arabidopsis thaliana*, to drive trait stacking in a crop species, to engineer HIV-resistant human T cells and hematopoietic stem cells (HSCs), and to drive targeted integration in ES cells, iPS cells, and mesenchymal stem cells (Urnov et al., 2010). Cathomen and Joung (2008) assert that Cys_2-His_2 ZF domains are the most abundant DNA-binding motif in eukaryotes.

Also, ZFNs can be used to edit genomes of numerous cell types and organisms. They induce site-specific DSB in DNA. Zinc finger recognition is dependent on matching the DNA sequence and mechanisms of DNA repair. This includes NHEJ and HR, which are shared by all species (Carroll, 2011). NHEJ-mediated targeting using ZFNs has been reported in mammalian tissue culture cells and *Arabidopsis*. In *Arabidopsis*, it has been reported with a lower frequency. In plant somatic cells, the primary mechanism of DSB repair is NHEJ. The use of NHEJ for targeting expands the range of explants and tissues that are available for genome editing using ZFNs (Bonawitz et al., 2019).

Efficiency

ZFNs have gene modification efficiencies of around 20% (Wang et al., 2017; Händel et al., 2009). These can be controlled using the DSB repair approach. DSB repair following HDR is less active than NHEJ in most of the cell cycle. This makes gene correction and insertion more challenging. However, delivering a homology donor DNA with ZFNs increases the rate of HDR at the DSB sites. ZFN–donor combination is thus used to improve the efficiency of gene modification using ZFNs in mammalian cells and disease treatment. Additionally, co-delivery of ZFNs with homology donor DNA achieves user-designed gene editing following the HDR repair pathway.

ZFNs can be used in clinical applications to achieve therapeutic editing. They are delivered to target cells. This can be done in vivo or ex vivo. This has the potential to treat some diseases that affect multiple organ systems. The barriers for this therapy include poor choices of vectors for transformation, immune responses against the ZFNs, and off-target mutations. During ex vivo therapy, target cells have to be removed from the body, modified with ZFNs, and then transfused into the host (Huang et al., 2018).

Specificity

ZFNs high specificity is because of the dimerization requirement for the FokI nuclease domain in the creation of a DSB. It requires that two ZFN monomers orient themselves with appropriate spacing at the target site. They can bind to any DNA sequence (Ran et al., 2018; Huang et al., 2018). This region consists of C_2H_2-zinc fingers. Each of them recognizes a 3 bp segment of DNA. Thus a ZFN has three zinc fingers that bind to 9 bp, thereby targeting DNA (Huang et al., 2018; Abbehausen, 2019). The "three-finger" ZFNs have little specificity and activity. The specificity can be improved by using up to six fingers per ZFN (Huang et al., 2018). It can also be improved by using FokI variants that require heterodimerization between the ZFN monomers. This also leads to a reduction of off-site cleavage that is due to the homodimerization of the individual ZFN monomers (Ran et al., 2018; Huang et al., 2018).

Engineered ZF domains with a modular assembly are about 100-fold more specific (three-finger context) than naturally occurring ones (Kim and Kini, 2017; Paschon et al., 2019). The design of ZFNs is usually modular and has thus been used in applications such as transposases, nucleases, integrases, recombinases, and gene regulators (Kim and Kini, 2017; Paschon et al., 2019). Zinc fingers occur in tandem. This gives them a binding specificity for DNA or RNA sequences of varying lengths (Abbehausen, 2019).

There are several efforts in coming up with approaches to enhance ZFNs' specificity. One of the approaches undertaken is the creation of obligate heterodimeric ZFN architectures that rely on charge–charge repulsion to prevent unwanted homodimerization of the FokI cleavage domain (Gaj et al., 2016). Another approach for improving ZFN specificity is to deliver them into cells as protein. According to Gaj et al. (2016), ZFN proteins possess inherent/intrinsic cell-penetrating activity, which enables them to facilitate cell editing with fewer off-target effects when applied directly onto cells as purified protein compared to when expressed within cells from nucleic acids.

The specificity of the engineered ZFNs can also be improved by;

1. Improving DNA-binding specificities of ZF domains.
2. Optimization of the linker sequence which connects the ZF domain with the FokI cleavage domain.
3. Regulation of DNA-cleavage activity of the FokI nuclease domain.

High ZFN specificity has been accomplished by the combination of two techniques (Urnov et al., 2010). The first of these is the usage of two-finger modules to assemble ZFPs with longer (12–18 bp) DNA recognition positions. Such elongated sites are potentially unusual even in complex genomes. However, it should be noted that a longer target site is indispensable but not adequate for enhanced specificity. Also, the protein and DNA must interact at the correct positions across the entire recognition interface to ensure efficient binding in vivo; one approach is to examine an archive of both designed and selected zinc finger modules and use data from such analysis to substitute certain residues or even whole α-helices in the ZFP with those that are probable to perform better in vivo. An example of such specificity is provided by the paralogue-specific action of the ZFNs designed to target the inositol-pentakisphosphate 2-kinase (Ipk1) locus in maize. The second, complementary approach is to require that a DSB only be prompted via a heterodimer of two ZFNs (via the use of obligate heterodimerization domains). ZFNs combining both approaches recognize composite sites of 24–36 bp (unique within the genome) and are well-tolerated by primary and transformed mammalian cells, zebrafish embryos, and rat embryos.

Monitoring Editing Specificity

Several independent assays have been developed to analyze the ZFN action genome-wide (Urnov et al., 2010). These are broadly divided into methods guided by a biochemical determination of the specificity of the two ZFP DNA-binding domains that incorporate a specified ZFN pair and approaches that are autonomous of such preceding information. Determining the specificity for a ZFP DNA-binding domain in vitro is done by systematic evolution of ligands by exponential enrichment (SELEX). This results in an experimentally determined unanimity binding position. A SELEX protocol has been developed that produces a biologically appropriate consensus for naturally occurring C_2H_2 ZFPs. The genome-wide specificity of ZFN action can be investigated by direct sequencing of these loci in cells already known to carry the wished-for modification at the targeted endogenous gene.

Error Rate

Gene knockouts occur when DSB repair follows the NHEJ pathway due to errors. The errors are due to indels in the break site that may occur because the NHEJ pathway is prone to such errors (Huang et al., 2018). Frameshift mutations in the target gene can disrupt it. This can potentially lead to the expression of the truncated and/or nonfunctional protein. This can result in cell or organism death (Carroll, 2011). This is because excess cleavage at off-target sites can produce breaks that outstrip the cell's DNA repair capacity. This shows that ZF DNA-binding domains have imperfect target-site recognition. Some of the types of zinc finger proteins are listed in Table 2.1.

Antibodies that are specific for p53 tumor suppressor-binding protein 1 (53BP1) or phosphorylated histone H2AX (γ-H2AX) can be used to quantify the cytotoxicity that is associated with ZFN expression. The assay characterizes the immediate genotoxicity and specificity of artificial nucleases of interest. However, they fail to provide information about off-target DSBs sites.

Versions of ZFNs

Non-classical types of ZFs differ in cysteine/histidine combinations. These include C2–CH, C2–H2, and C2–C2. There are about 30 types of ZFNs that are approved by the HUGO Gene Nomenclature Committee. The classification is based on the structure of the zinc finger domain. C2–H2 domain proteins are among the most important and abundant. Of interest are the new gene (RING), plant homeodomain (PHD), Lin-II, Isl-1, and Mec-3 (LIM domains). Protein structures of the domains are presented in Table 2.1. Several transcription factors have the C-x-C-x-H-x-H motif among C2-H2 ZNFs. This motif mediates the direct interaction of the ZFN with DNA. ZNF217, which is among the C2-H2 members, contains many C2-H2 domains that bind to specific DNA sequences such as (T/A)(G/A)CAGAA(T/G/C). This represses expression of the target genes (Cassandri et al., 2017).

Construct

The zinc finger is a platform for the design of innovative DNA-binding domains (Urnov et al., 2010). The ZFP region avails a ZFN with the capability to bind a distinct base sequence. This region comprises a tandem array of Cys_2-His_2, each distinguishing approximately 3 bp of DNA. There are a variety of strategies for creating ZFPs with new, user-chosen binding specificities. The first arose from observations of the first ZFP–DNA co-crystal structure, which suggested a substantial degree of functional independence in the interaction of discrete fingers with DNA. Another approach is termed "modular assembly". This approach generates candidate ZFPs for a specified target sequence by detecting fingers for each triplet component and connecting them into a multi-finger peptide targeted to the corresponding compound sequence. Urnov et al. (2010) state that fingers used for modular assembly have been created for most triplet sequences. According to Urnov et al. (2010), the method has been used to develop the zinc finger component of active ZFNs for numerous endogenous targets in higher eukaryotic cells.

In addition to modular assembly, several alternative approaches for building ZFPs have been developed. These newer methods were designed to cater to the deviations from stringent functional modularity witnessed for many zinc fingers. It has been noted that many designed and natural fingers can contact contiguous fingers as well as bases external to their proximal DNA triplet. Although such interactions can allow more selective binding, they can also obfuscate efforts to create new ZFPs through modular design.

Whatever the design method, the creation of a DNA-binding module assessed in vitro for affinity and specificity in the direction of its projected target provides only the initial step toward use in vivo. ZFNs assembled from in vitro "validated" ZFPs normally fail to drive genome editing at the endogenous locus when tested in living cells. Complex genomes often contain naturally occurring multiple copies of a sequence that is similar to the intended target (for example,

Table 2.1 Types of Zinc Finger Proteins

Type name	Zinc finger structure	Number of genes	Number of TF	Important members
Zinc fingers XE "fingers" C2H2-type (ZNF)	C-x-C-x-H-x-H	720	372	*KLF4, KLF5, EGR3, ZFP637, SLUG, ZNF750, ZNF281, ZBP89, GLIS1, GLIS3*
Ring finger proteins XE "proteins" (RNF)	C-x-C-x-C-x-H-xxx-C-x-C-x-C-x-C	275	12	*MDM2, BRCA1, ZNF179*
PHD finger proteins XE "proteins" (PHF)	C-x-C-x-C-x-C-xxx-H-x-C-x-C-x-C	90	0	*KDM2A, PHF1, ING1*
LIM domain containing	C-x-C-x-H-x-C-x-C-x-C-x-C-x-(C, H, D)	53	1	*ZNF185, LIMK1, PXN*
Nuclear hormone receptors (NR)	C-x-C-x-C-x-C-xxx-C-x-C-x-C-x-C	50	47	*VDR, ESR1, NR4A1*
Zinc fingers XE "fingers" CCCH-type (ZC3H)	C-x-C-x-C-x-H	35	2	*RC3H1, HELZ, MBNL1, ZFP36, ZFP36L1*
Zinc fingers XE "fingers" FYVE-type (ZFYVE)	C-x-C-x-C-x-C-xxx-C-x-C-x-C-x-C	31	0	*EEA1, HGS, PIKFYVE*
Zinc fingers XE "fingers" CCHC-type (ZCCHC)	C-x-C-x-H-x-C	25	2	*CNBP, SF1, LIN28A*
Zinc fingers XE "fingers" DHHC-type (ZDHHC)	C-x-C-x-H-x-C-xxx-C-x-C-x-H-x-C	24	0	*ZDHHC2, ZDHHC8, ZDHHC9*
Zinc fingers XE "fingers" MYND-type (ZMYND)	C-x-C-x-C-x-C-xxx-C-x-C-x-H-x-C	21	4	*PDCD2, RUNX1T1, SMYD2, SMYD1*
Zinc fingers XE "fingers" RANBP2-type (ZRANB)	C-x-C-x-C-x-C	21	3	*YAF2, SHARPIN, EWSR1*
Zinc fingers XE "fingers" ZZ-type (ZZZ)	C-x-C-x-C-x-C	18	3	*HERC2, NBR1, CREBBP*
Zinc fingers XE "fingers" C2HC-type (ZC2HC)	C-x-C-x-H-x-C	16	2	*IKBKG, L3MBTL1, ZNF746*
GATA zinc finger domain containing (GATAD)	C-x-C-x-C-x-C	15	15	*GATA4, GATA6, MTA1*
ZF class homeoboxes and pseudogenes	C-x-C-x-H-x-H	15	10	*ADNP, ZEB1, ZHX1*
THAP domain-containing (THAP)	C-x-C-x-C-x-H	12	3	*THAP1, THAP4, THAP11*
Zinc fingers XE "fingers" CXXC-type (CXXC)	C-x-C-x-C-x-C-xxx-C-x-C-x-C-x-C	12	2	*CXXC1, CXXC5, MBD1, DNMT1*

(Continued)

Table 2.1 Types of Zinc Finger Proteins (Continued)

Type name	Zinc finger structure	Number of genes	Number of TF	Important members
Zinc fingers XE "fingers" SWIM-type (ZSWIM)	C-x-C-x-C-x-H	9	0	*MAP3K1, ZSWIM5, ZSWIM6*
Zinc fingers XE "fingers" AN1-type (ZFAND)	C-x-C-x-C-x-C-xxx-C-x-H-x-H-x-C	8	0	*ZFAND3, ZFAND6, IGHMBP2*
Zinc fingers XE "fingers" 3CxxC-type (Z3CXXC)	C-x-C-x-H-x-C	8	0	*ZAR1, RTP1, RTP4*
Zinc fingers XE "fingers" CW-type (ZCW)	C-x-C-x-C-x-C	7	0	*MORC1, ZCWPW1, KDM1B*
Zinc fingers XE "fingers" GRF-type (ZGRF)	C-x-C-x-C-x-C	7	0	*TTF2, NEIL3, TOP3A*
Zinc fingers XE "fingers" MIZ-type (ZMIZ)	C-x-C-x-H-x-C	7	1	*PIAS1, PIAS3, PIAS4*
Zinc fingers XE "fingers" BED-type (ZBED)	C-x-C-x-H-x-H	6	2	*ZBED1, ZBED4, ZBED6*
Zinc fingers XE "fingers" HIT-type (ZNHIT)	C-x-C-x-C-x-C-xxx-C-x-C-x-H-x-C	6	0	*ZNHIT3, DDX59, INO80B*
Zinc fingers XE "fingers" MYM-type (ZMYM)	C-x-C-x-C-x-C	6	6	*ZMYM2, ZMYM3, ZMYM4*
Zinc fingers XE "fingers" matrin-type (ZMAT)	C-x-C-x-H-x-H	5	0	*ZNF638, ZMAT1, ZMAT3, ZMAT5*
Zinc fingers XE "fingers" C2H2C-type	C-x-C-x-H-x-H	3	3	*MYT1, MYT1L, ST18*
Zinc fingers XE "fingers" DBF-type (ZDBF)	C-x-C-x-H-x-H	3	0	*DBF4, DBF4B, ZDBF2*
Zinc fingers XE "fingers" PARP-type	C-x-C-x-H-x-C	2	1	*LIG3, PARP1*

Source: Cassandri et al., 2017.

paralogues, or pseudogenes), and these copies can act as additional targets for ZFNs. Another problem is the chromatin structure at target sites, which may not be acquiescent to cleavage. This can be overcome by covering the target region with a large ZFN panel, followed by screening for activity directly at the endogenous locus and, when necessary, by iteratively optimizing the protein DNA interface.

The FokI domain has been fundamental to the success of ZFNs, as it possesses several characteristics that support the goal of targeted cleavage within complex genomes. FokI must dimerize to cleave DNA. As this interaction is weak, cleavage by FokI as part of a ZFN requires two contiguous and autonomous binding events, which must occur in both the accurate location and with proper spacing to permit dimer formation. The requirement for two DNA-binding events allows specific targeting of elongated and potentially inimitable recognition sites (from 18–36 bp). Furthermore, the reliance on productive dimerization has spurred the development of variants that cleave only as a heterodimer pair, thus enhancing specificity via the elimination of undesirable homodimers. In some ongoing efforts, the ZFP–Fok linker has been altered as a means of developing ZFN dimers with novel spacing requirements for the two monomer binding events. FokI catalytic domain variants with improved cleavage activities have also been described (Urnov et al., 2010).

In eukaryotes, the most abundant motifs are the Cys_2 -His_2 ZF domains. They consist of ~30 residues. These residues fold into a $\beta\beta\alpha$-structure. This is coordinated by a zinc ion (Cassandri et al., 2017). An array of synthetic ZF domains that can range from three to six assists in the DNA binding of ZFNs. The domains have a corresponding 12–18 bp binding site. The endonuclease domain of FokI provides nuclease activity. ZFNs are normally used in pairs. One member of each pair binds on either side of the target site. Off-target endonuclease cleavage and the self-association of monomers are reduced by using two obligately heterodimerizing "high-fidelity" ("HiFi") FokI domains on the ZFN monomers (Bonawitz et al., 2019). How ZFNs bind to DNA is illustrated in Figure 2.1.

ZFNs consist of two functional domains. The customized ZF array specifies DNA binding while the endonuclease domain of FokI has catalytic activity (Zahur et al., 2017). Binding to the target half-site is conferred by the ZF array in the ZFN subunit. The dimerization of the two subunits occurs in the correct orientation and spacing. The DSB is introduced by the nuclease domain. It is introduced within the spacer sequence separating the target half-sites (Carroll, 2011; Zahur et al., 2017). The DSBs are then repaired via NHEJ. This process is mediated by the cell repair machinery. Small indels can occur leading production of nonfunctional protein products. The generation of large nonfunctional protein moieties can be prevented by designing ZFNs that target exons in mRNA (Zahur et al., 2017). The molecular functions, structure and the subcellular localization of ZNFs are shown in Figure 2.2.

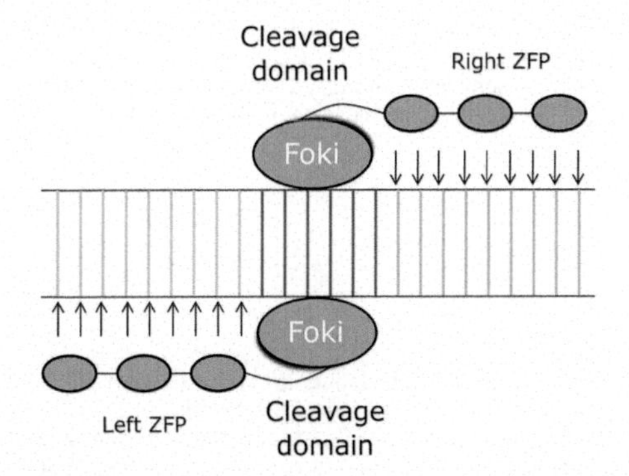

Figure 2.1 The illustration shows how a pair of ZFNs bind to DNA. Schematic illustration of zinc finger nuclease (ZFN) structure and mechanism of inducing DSBs on its target. The target site of the ZFN is recognized by the "left" and "right" monomers consisting of a tandem array of three to six engineered zinc finger proteins (ZFPs) (three are shown here); a single engineered ZFP can recognize a nucleotide triplet. Each ZNF is linked to a nuclease domain from the FokI restriction enzyme. Recognition of the target sequence by the left and right ZFPs results in dimerization of the FokI nuclease; DNA cleavage takes place along the spacer sequence (usually 6 bp long, shown in dark grey) between the two ZFP recognition sites.

Source: Limera et al., 2017.

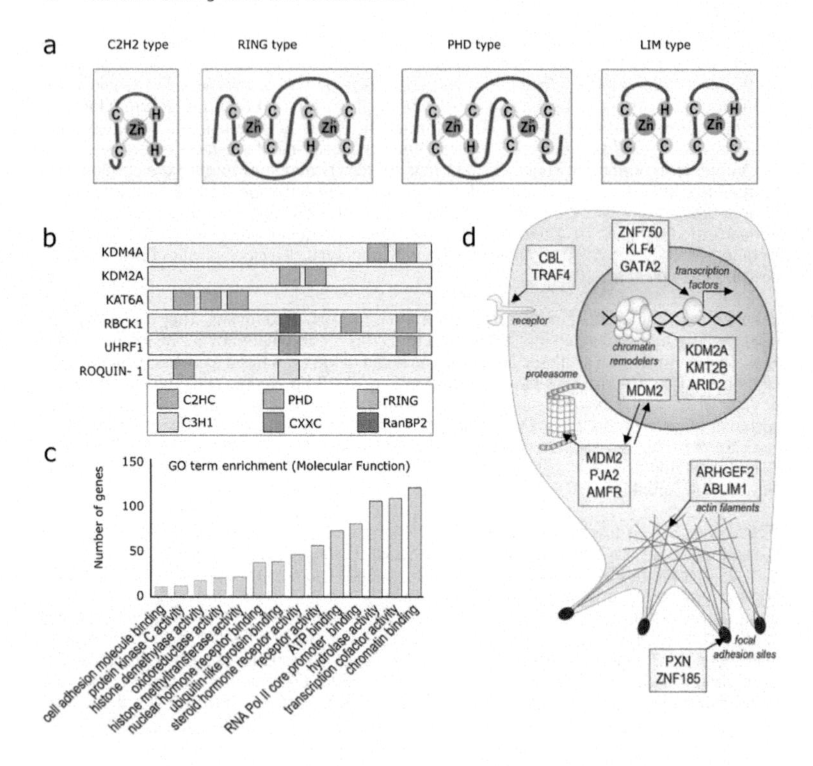

Figure 2.2 Shows the molecular functions, structure, and the subcellular localization of ZFNs. (a) The structure of C2H2, RING, PHD, and LIM ZF domains. (b) The structure of ZFNs that have multiple ZF domains. (c) The gene ontology analysis of 1,723 annotated ZFNs according to their molecular function. (d) The subcellular localization of different ZFNs.

Source: Cassandri et al., 2017.

Procedure

ZFNs create DSB in the target site, stimulating cell repair (Miki et al., 2021). A prokaryotic type IIS restriction endonuclease from bacteria, FokI, recognizes the non-palindromic penta-deoxyribonucleotide 5'-GGATG-3':5'-CATCC-3' that is found in duplex DNA. It cleaves 9/13 nt downstream of its recognition site (Durai et al., 2005). However, it does not recognize a specific sequence at the cleavage site. This shows that there are two separate protein domains in FokI. One is for sequence-specific DNA recognition. The other is for endonuclease activity. A signal is sent to the endonuclease domain using allosteric interactions. This occurs after anchoring the DNA-binding domain at the recognition site. The cleavage of the DNA consequently occurs. It is possible to swap natural DNA-binding proteins with longer DNA recognition sequences with the FokI recognition domain to produce chimeric nucleases. The swap can be done with synthetic DNA-binding motifs (Durai et al., 2005). The release of the endonuclease domain from the recognition domain allows it to swing over the DNA cut site. DNA cleavage occurs only when FokI is bound to its cognate site. This occurs only when magnesium ions (Mg^{2+}) are present. The dimerization of the endonuclease domain is essential for FokI to produce a DSB (Durai et al., 2005). Stages during DNA recognition and cleavage by ZFNs are shown in Figure 2.3.

ZFN-Mediated Genome Editing

A DSB creating ZFN has two monomer subunits. Individual subunits each have three ZFs. These recognize 9 bp in the target site and the FokI endonuclease domain. The two domains are connected by a short linker. The nuclease is activated after dimerization. After activation, it can cut the DNA in the spacer sequence. This separates the target half-sites (Left) and (Right).

A DSB introduced by the ZFN is a dominant mutant allele. This is repaired following the NHEJ pathway. Indels disrupt the coding sequence, making the protein product nonfunctional.

In order to restore a genetic defect directly in the genome, a targeting vector (donor DNA) is transduced into the cell that is targeted. It should encompass the wild-type sequences homologous to the mutant gene. This allows for the correction of a genetic defect in the genetic material. The DSB induced by ZFN stimulates the HR repair pathway. This occurs between donor DNA and the defective gene. It generates a corrected locus.

A partial cDNA is flanked by sequences that are homologous to the mutant gene put in the targeting vector. HR repair pathway is stimulated by the ZFN-induced DSB. HR will occur between donor DNA and the mutant gene. The expression of the gene is reconstituted. It will remain under the control of the endogenous promoter. This restores the phenotype of a cell that harbors a genetic defect.

Applications of ZFNs

The applications of genome editing employing ZFNs are grounded on the introduction of a site-specific DNA DSB into the locus of concern (Urnov et al., 2010). All eukaryotic cells efficiently repair DSBs via the homology-directed repair HDR also referred to as HR or nonhomologous end joining (NHEJ) pathways. These highly conserved pathways can be capitalized on to generate distinct genetic endings across an extensive range of cell types and species. NHEJ repair, for example, rapidly and efficiently ligates the two broken ends, with the occasional gain or loss of genetic material. It can thus be used to present small insertions and/or deletions at the site of the break, an outcome that can be exploited to interrupt a target gene. Some of the applications of ZFNs are discussed in the following sections.

Gene Disruption

The unpretentious way of genome editing, gene disruption, takes advantage of errors introduced during DNA repair to obliterate the function of a gene or genomic region. This approach has been applied in innumerable species and cellular contexts to knock out user-specified genes in a single step and starved of selection for the sought-after event. Examples are provided next.

Gene Disruption in Model Organisms

To disrupt a gene in *D. melanogaster*, ZFNs targeting exonic sequences can be delivered via mRNA injection into the early fly embryo. Up to 10% of the progeny formed by the resulting adult flies is mutated for the gene of concern. For one gene (coilin), six different alleles were acquired, all containing small frameshift-inducing deletions. Animals homozygous for each allele were devoid of expression of the protein encoded by the target gene. Injection of mRNA encoding engineered ZFNs into embryos has also been used to generate zebrafish containing preferred genetic lesions. In four distinct studies, up to 50% germline mosaicism at the targeted genes were obtained. For gene disruption in rats, in which early development takes place much slower than in insects or fish, engineered ZFNs with extended recognition sites were used. This created knockout animals for two discrete endogenous genes, and ZFN-treated founders transmitted disrupted alleles at a frequency of 10–100%. A rat model of severe combined immune

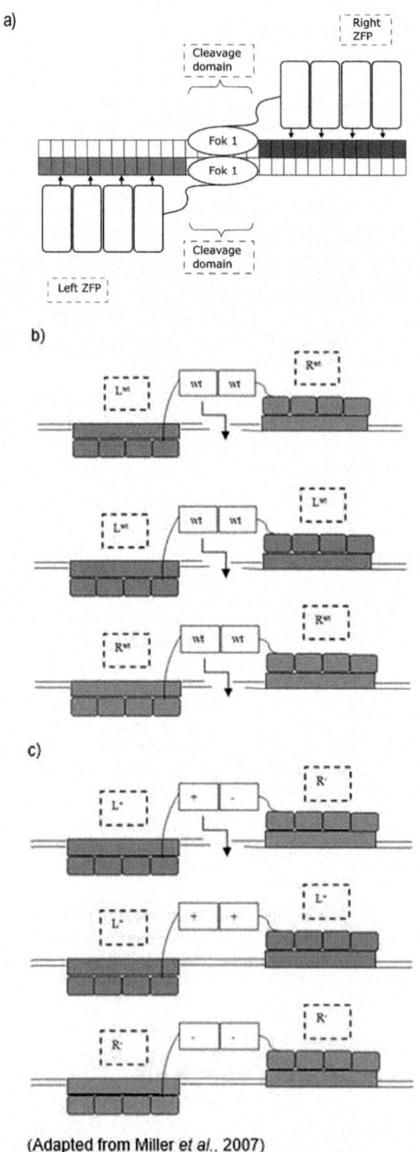

Figure 2.3 Stages during DNA recognition and cleavage by ZFNs.

(a) ZFN dimer is bound to a nonpalindromic DNA target. Each ZFN has the cleavage domain of FokI that is fused to a zinc finger protein (ZFP) customized to specifically recognize a "left" or "right" half-site (shown by dark and light gray boxes). These sites are separated by a spacer of 5 or 6 bp. The simultaneous binding of both ZFNs allows for dimerization of the FokI nuclease domain and cleavage of DNA. Endogenously active ZFNs can be created using four fingers (indicated here); each binds 12 bp sites. They can have three fingers, each binding 9 bp sites.

deficiency (SCID) was also produced. In systems in which mRNA microinjection is currently not an option (for example, the model plant *A. thaliana*), stable transgenesis of an inducible ZFN expression cassette allows gene disruption. As the path to a knockout organism is now one generation-long, ZFN-driven gene disruption, even when allowing for a period of ZFN development, equates satisfactorily in terms of period and screening effort with other approaches for producing targeted knockouts (such as classical gene targeting in mouse ES cells).

Gene Disruption in Mammalian Somatic Cells

ZFN-driven gene disruption has also been used for mammalian somatic cell genetics, in which the ZFN is transiently expressed followed by an analysis of single cell-derived clones. Classical gene targeting combined with positive and negative selection strategies is a powerful tool for gene knockout in mouse ES cells, and the use of engineered adeno-associated viruses (AAVs) has permitted its application in transformed and primary human cells. ZFNs obviate the need for drug selection, extend the application of gene knockout to potentially any cell type and species for which transient DNA or mRNA delivery is available, and result in knockouts in 1–50% of all cells. In recent times, transient hypothermia has been shown to further escalate ZFN-driven gene disruption regularity in transformed and primary cells by two- to fivefold. The first published case in point of the use of engineered ZFNs to disrupt an endogenous locus in a mammalian cell involved a knockout of the dihydrofolate reductase (DHFr) gene in Chinese hamster ovary (CHO) cells. A plasmid encoding the ZFNs was made known by transient transfection, which resulted in disruption frequencies of up to 15% of alleles in the cell population. Limiting dilution and genotyping yielded two clones (out of ~60) in which the Dhfr gene was biallelically disrupted and lacked measurable DHFr protein expression. Subsequently, the sequential or simultaneous application of locus-specific ZFNs has been used to efficiently make double and triple locus gene knockouts in CHO and K562 cells.

More Complex Types of Disruption

The array of mutations that can be produced via NHEJ need not be restricted to small insertions or deletions at the ZFN target position. For example, ZFN cleavage at a proximal pair of positions at the same locus deletes the intervening fragment, and other studies have shown that more distal pairs of target sequences may be cleaved and joined, even though with decreased adeptness. Finally, in both hamster and human cells, supplying the cells with a double-stranded oligonucleotide carrying overhangs complementary to those that the ZFNs generate in the endogenous locus produces chromatids in which the oligonucleotide has precisely ligated into the chromosome. This approach can be used to add tags to endogenous genes in cells in which HDR is less efficient or to substitute an entire chromosomal section with a recombinase recognition site. With some notable exceptions, mammalian somatic cell "genetics" has required quotation marks, as it relied on RNAi. This is a process that offers an operationally beneficial ephemeral knockdown of target gene expression but not an authentic genetic modification. As

Figure 2.3 (Continued)

(b) Co-expression of Lwt (left ZFP fused to the wild-type FokI cleavage domain) and Rwt (right ZFP fused to wild-type FokI cleavage domain). This yields a heterodimer that cleaves the target sequence (top). It also yields homodimers that can cleave other targets (middle and bottom). Self-complementary polygons (wt) are the wild-type FokI cleavage domains. Four adjacent boxes are ZFP domains. The capacity of individual dimmers to cleave a DNA sequence is shown by jagged arrows.

(c) The solution is offered by the new protein designs. The variants of the FokI cleavage domain function as obligate heterodimers (complementary polygons labeled "+" and "–"). They enable cleavage of the desired heterodimer targets (top). They also eliminate off-target activity caused by homodimer species (middle and bottom). L+ and R– denote the "left" and "right" ZFPs, respectively. These are fused to variant cleavage domains. The absence of a jagged arrow that has the L+/L+ and R–/R– pairings shows the inability of the homodimers to cleave DNA.

larger numbers of functionally validated gene-specific ZFNs become available, a gene knockout will become the standard for determining gene function.

Homology-Based Genome Editing

HDR can be invoked following a ZFN induced DSB as alluded to earlier. Whether impromptu or prompted by the I-SceI homing endonuclease or a ZFN, a DSB is recombinogenic in cells of higher eukaryotes. Homology-based genome editing requires the concurrent provision of a suitably designed, homology-containing donor DNA molecule along with the locus-specific ZFNs. This enables two related modes of genome editing that are a function of donor DNA design and distinguished by the type of allele being generated.

Gene Correction (Allele Editing)

This approach allows the transmission of single-nucleotide changes and short heterologous expanses from an episomal donor to the chromosome, succeeding in a ZFN-induced DSB. The latest experiments support the view that the endogenous repair machinery uses the extrachromosomal, investigator-provided donor as a template for repairing the DSB via the synthesis-dependent strand annealing process. This technique enables the study of gene function and/or the modeling of disease-causing mutations through the creation of a point mutation that is characteristic, for example, of a known disease-predisposing allele or that disables a motif that is thought to be crucial for function. Such point mutations can be efficiently created at a specific position in the target gene.

Gene Addition in Human ES and iPS Cells

The application of this method in human mesenchymal stem cells generated 50% targeted gene accumulation deprived of selection. Furthermore, ZFNs delivered as plasmid DNA have been used in human ES and iPS cells to efficiently target a drug resistance marker to a precise gene and to generate novel allelic forms of three endogenous loci. In all of these studies, efficient, specific, and stable gene addition was achieved, and the cells retained characteristics of pluripotency.

Gene Addition in Plants

ZFN-driven gene addition to native loci has of late also been accomplished in plants. This is a taxonomic group that historically has proven resistant to targeted gene modification. ZFNs targeting an endogenous endochitinase gene in tobacco were co-delivered to tobacco suspension cells or leaf disk protoplasts along with a donor DNA carrying short homology arms and a herbicide resistance marker. According to Urnov et al. (2010), the correct addition of the resistance cassette was observed in ~10% of the cases. A recent report described the editing of two endogenous loci in *Zea mays* (maize). It is reported that ZFN-edited plants were fertile, the transgene was transmitted to the next generation in normal Mendelian fashion, and co-segregation of the herbicide-resistant and low-phytate-content phenotype was observed. Site-specific gene addition in a major crop species could be used for "trait stacking", an important goal that involves the creation of plants in which several independent traits are physically linked, thus ensuring their co-segregation throughout the breeding process (Urnov et al., 2010).

ZFNs can be applied in functional genomics/target validation. They can be used to create gene knockouts in multiple cell lines and completely knockout genes that are not amenable to RNAi. They can also be used in cell-based screening. They can also be used to create knock-in cell lines that have the promoters, reporters, or fusion tags integrated into the endogenous genes. ZFNs can create cell lines that can produce higher yields of proteins and or antibodies.

Cathomen and Joung (2008) state that by applying the ZFN technology to stem cells, inherited mutations may be fixed ex vivo, and functionally rectified stem cells could be transplanted back into patients to repopulate the affected tissues and remedy the disease. Importantly, gene correction would give back the functionality of the affected DNA segment and, at the same time, restore its normal endogenous expression pattern, in so doing asphyxiating a major constraint of conventional gene therapy methodologies. Gene rectification can work even more efficiently if the fixed gene provides the altered stem cell with a growth benefit. In the case of X-linked severe combined immunodeficiency, which is triggered by mutations in the IL2Rγ locus, the rectification of only a rarity of genetically rectified HSCs will be ample to reinstate the proper function of the immune system. Several impediments carry on to restrict the utilization of ZFNs in a therapeutic setting.

According to Cathomen and Joung (2008), the following criteria must be achieved for ZFNs to be successfully applied in a clinical setting:

1. High DNA-binding specificity of the ZF domain.
2. Regulated cleavage by the ZFN.
3. Efficient delivery.
4. Transient ZFN expression.
5. A comprehensive evaluation of treated cells for potential ZFN-induced side effects.
6. Assessment of the potential immune reactivity against ZFNs, especially against the bacterial FokI domain.

Other Practical Applications of ZFNs

ZFNs were used to efficiently and specifically excise HIV-1 proviral DNA. This was achieved in latently infected human T cells. To achieve this, long terminal repeats (LTRs) were targeted. This is an alternative and novel antiretroviral strategy that can be used in eradicating HIV-1 infection

Table 2.2 Reported Instances of Successful ZFN-Induced Gene Targeting

Organism	Latin name	Method	TM	TGR
Fruit fly	*Drosophila melanogaster*	Heat-shock induction	+	+
		Embryo injection	+	+
Nematode	*C. elegans*	Gonad injection	+	
Silkworm	*Bombyx mori*	Embryo injection	+	
Zebrafish	*Danio rerio*	Zygote injection	+	
Sea urchin	*Hemicentrotus pulcherrimus*	Embryo injection	+	
Frog	*Xenopus tropicalis*	Embryo injection	+	
Rat	*Rattus norvegicus*	Zygote injection	+	+
Mouse	*Mus musculus*	Zygote injection	+	+
Cress	*A. thaliana*	Agrobacterium	+	
Tobacco	*Nicotiana* sp.	Protoplasts	+	+
		Agrobacterium	+	+
		Viral delivery	+	
Maize	*Zea mays*	Cell culture	+	+
Petunia	*Petunia* sp.	Viral delivery	+	
Human	*Homo sapiens*	DNA XE "DNA" transformation	+	+
		Viral delivery	+	+
Mouse	*M. musculus*	DNA XE "DNA" transformation	+	+
Hamster	*Cricetulus griseus*	DNA XE "DNA" transformation	+	+
Pig	*Sus domestica*	DNA XE "DNA" transformation	+	

Source: Carroll, 2011.
Note: TM is targeted mutagenesis by NHEJ. TGR is the targeted gene replacement following HR.

(Ji et al., 2018). Table 2.2 shows more examples of organisms whose genome was edited using ZFNs.

Advances in Therapy Using ZFNs

Due to its relative simplicity, ZFN-mediated gene disruption (achieved by transient delivery of the ZFNs alone) is the first ZFN-based method that has been taken to the clinic, specifically for the treatment of glioblastoma (NCT01082926) and HIV76, (NCT00842634 and NCT01044654). In the former case, the glucocorticoid receptor gene is disrupted by ZFNs as part of a T cell-based cancer immunotherapy (autologous T cell therapy for the treatment of prostate cancer was recently approved by the US Food and Drug Administration). The AIDS trials are based on the fact that HIV infection requires the co-receptors chemokine (C-C motif) receptor type 5 (CCr5) or chemokine (C-X-C motif) receptor type 4 (CXCr4). A naturally occurring human mutation in the CCR5 gene (CCR5Δ32) was shown to confer resistance to the virus without causing detectable pathophysiological effects beyond an increased susceptibility to West Nile virus infection, and reducing or blocking CCr5–HIV interaction is therefore a validated drug target for small molecule inhibitors, small interfering RNA knockdown approaches, antibodies, or intrabodies (Urnov et al., 2010).

Advantages of ZFNs

ZFN-stimulated gene targeting can preserve temporal and tissue-specific gene expression better than conventional gene-addition-type gene therapy. Gene knockout through NHEJ-mediated repair of ZFN-induced DSBs is another promising application of this technology. Gene disruption may also allow the treatment of hereditary disorders with dominant inheritance if the mutated allele can be disrupted specifically by ZFN-induced DSBs (Cathomen and Joung, 2008).

ZFNs achieve rapid disruption and integration of DNA sequences into any genomic loci. The mutations they make are permanent and can be heritable. They work in many different mammalian somatic cell types. The genome edits can be induced through one transfection experiment. They knock out or knock in cell lines in a short time of around two months. The single or bi-allelic edits can occur in 1–20% of the clone population, and there is no need for antibiotic selection when screening transformants.

Disadvantages of ZFNs

Off-target mutations are common when ZFNs are used. This is because the modularity may be somewhat unpredictable. ZFNs have complex requirements in designing and constructing them, resulting in a high failure rate. They have a limited application in high throughput screening.

The other disadvantage is that the necessity for co-delivery to the target cell of the donor DNA construct alongside with the ZFNs can often be restrictive and require optimization. The other issue is repaired pathway choice or competition, that is, the propensity of a cell to preferentially repair a ZFN-induced DSB using NHEJ rather than HDR Notable, disabling the ligase gene in *D. melanogaster* (the product which is required for NHEJ) strongly prejudices the resolution of a ZFN-induced break in the direction of HDR (Urnov et al., 2010).

3 TALENs

History

Transcription activator-like effector nucleases (TALENs) are synthetic molecules that are composed of transcription activator-like effectors (TALEs) and FokI endonucleases. They were discovered in the year 2007 (Congressional Research Service, 2018; DeFrancesco, 2011). TALENs have since been used in genome editing. They are easy to design and less expensive. They can be used to edit and up-regulate genes (Bloom et al., 2015). TALENs are similar in nature to ZFNs. They comprise of a nonspecific FokI nuclease domain fused to a customizable DNA-binding domain. The DNA-binding domain of TALENs is made of TALE domains (Joung and Sander, 2013). There are four different TALE domains, one for each DNA base, so they can be engineered to bind to specific DNA sequences much more easily than ZFNs. Like ZFNs, the nuclease part of TALENs is normally a FokI nuclease. Two FokI molecules must come together to cut the DNA, so two TALENs are made, one for each strand. When TALEs are fused to the FokI nuclease, they only cleave DNA when present as dimers (Bloom et al., 2015). TALENs, therefore, function in pairs. They achieve this by binding opposing DNA targets. This will be across a spacer over which the FokI domains come together to create a DSB in the DNA strand (Carter, 2016).

There are some transcription activator-like (TAL) effectors produced in plants. They are a family of virulence factors that are produced by plant pathogens (*Xanthomonas* spp.). They can be imported into nuclei, where they act as transcriptional activators (Min et al., 2018). This technology was patented and commercialized.

Principle

TALENs utilize repeating units of TALEs as a DNA recognition module and a FokI catalytic nuclease domain as the DNA cleavage module (Bloom et al., 2015). TALEs have repeating domains of between 33 and 35 amino acids. These units are largely identical except for two amino acids found at positions 12 and 13. These are referred to as the repeat variable diresidues (RVDs). They specifically recognize a single nucleotide. They thus provide DNA recognition specificity (Rode et al., 2019). The diamino acid/DNA recognition code together with its modular nature makes TALEs a good choice for creating targeted DNA nucleases. TALENs together with a desired exogenous DNA can facilitate targeted gene editing following HR in mammalian genomes (Prowse et al., 2017). Recombination rates in mammalian cells are around 105–106 events/cell/generation. This is a limitation of TALENs. Chromosomal DSB catalyzed by TALENs can stimulate HR by greater than 1,000-fold (Min et al., 2018).

Procedure

The mode of action for TALENs is similar to the one for ZFNs. Expressing a fusion protein made from the endonuclease FokI and an engineered DNA-binding domain that is derived from TAL proteins generates a DSB (Min et al., 2018; Petersen, 2017). Each monomer binds to a "half-site" in the target, and the FokI endonuclease domains dimerize to generate a DSB in the spacer sequence between the two half-sites. Dimerization of these fused domains enables the recognition of two half-sites, designated "right" and "left". The formation of a DSB is mediated by FOKI (Joung and Sander, 2013). These proteins are isolated from the plant pathogenic *Xanthomonas*, and they bind DNA because of a central domain that consists of about 34 amino acid tandem

DOI: 10.1201/9781003165316-3

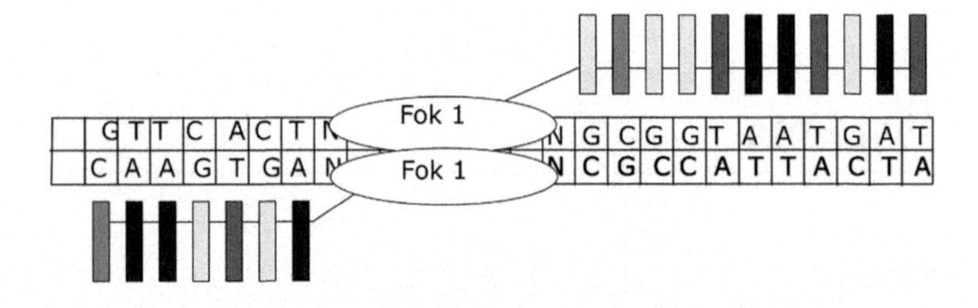

TALEs bind to nucleotides

A= C= G= T=

Figure 3.1 Illustration of TALENs. (Adapted from Usal et al., 2010.)

repeats (Macias et al., 2017). The identity of the amino acids found at positions 12 and 13 are hypervariable, the RVD (Carroll, 2017). They are responsible for the specificity of amino acid recognition (Bauman et al., 2018; Feng et al., 2014). Increasing or decreasing the number of repeats in a TALEN enzyme affects the specificity of the recognition of the DNA sequence target. The construction of highly specific DNA-binding domains, therefore, is achieved by combining several RVDs (Moore et al., 2012). This engineering of the DNA-binding domain offers an advantage over ZFNs, as ZFNs in arrays will influence the specificity of adjacent ZFs. Therefore, the specificity of such arrays is difficult to determine. TALENs circumvent this problem as the DNA-binding specificity displayed by each TALE domain is independent. TALE domains can be extended to produce arrays, which can recognize lengths of any size. The mode of action of TALENs is illustrated in Figure 3.1.

Efficiency

TALENs are moderately inexpensive and difficult to engineer (AAS, 2017). TALENs have a significantly higher rate of cleavage activity as compared to ZFNs. They also have an essentially limitless targeting range and are suitable for use by nonspecialist researchers (Pu et al., 2015). They recognize single nucleotides (Liddicoat, 2016). Engineering interactions between TALENs DNA-binding domains and target nucleotides are straightforward (Kruminis-Kaszkiel et al., 2018). They are engineered using a simple "protein-DNA code" that relates modular DNA-binding TALE repeat domains to individual bases in a target-binding site (Joung and Sander, 2013). It is far more straightforward to engineer interactions between TALENs DNA-binding domains and their target nucleotides than it is to create interactions with ZFNs and their target nucleotide triplets. TALENs are about 200 times more expensive than some genome editing tools, and they take longer to perform (Schaap, 2018). Despite this fact, TALENs have been effectively used to modify endogenous genes in yeast, fruit fly, roundworm, cricket, zebrafish, frog, rat, pig, cow, thale cress, rice, silkworm, and human somatic and pluripotent stem cells (Feng et al., 2014). This of course was before the discovery and invention of CRISPR technology.

TALENs can have off-target activity leading to unwanted DSB. This can yield chromosomal rearrangements and or cell death (Rode et al., 2019). Studies have been carried out to compare the relative nuclease-associated toxicity of available technologies. TALENs constructs have a high precision based on the maximal theoretical distance between DNA binding and nuclease activity (Zhang et al., 2011). TALENS are more efficient than ZFNs but less efficient than CRISPR. TALEN efficiency can be improved by optimization of the TALE portion.

Applications of TALENs

Most uses of TALENs have so far been in scientific research, for example, to investigate models of human disease. However, the potential application of TALENs is much wider than just research, given that TALENs have the potential to alter any DNA sequence, whether in a bacterium, plant, animal, or human beings. They have an almost limitless range of possible applications in living things. Areas of research and possible applications include crops and livestock (e.g., increasing yield, introducing resistance to disease and pests, tolerance of different environmental conditions); industrial biotechnology (e.g., developing "third generation" biofuels and producing chemicals, materials, and pharmaceuticals); biomedicine (e.g., pharmaceutical development, xenotransplantation, gene and cell-based therapies, control of insect-borne diseases); and reproduction (e.g., preventing the inheritance of a disease trait) (Molina et al., 2011).

Currently, most research based on TALENs tool editing is done to understand diseases using cells and animal models (Gaj et al., 2013). Scientists are still working to determine whether this approach is safe and effective for use in people (Joung and Sander, 2013).

TALENs have been used to modify plant genomes to improve characteristics such as nutritional qualities (Carter, 2016). They have also been used to develop biofuels (Liddicoat, 2016). TALENs have been used to engineer human ES cells, iPSCs, and human erythroid cell lines, and to knock out genes in *C. elegans*, rats, mice, and zebrafish (Moore et al., 2012; Goold et al., 2018). Additionally, genes have been knocked out in several organisms such as cattle and rats by TALEN mRNA microinjection in one-cell embryos (Pu et al., 2015).

TALENs can be used in vitro to correct genetic defects that cause sickle cell disease, xeroderma pigmentosum, and epidermolysis bullosa (Congressional Research Service, 2018). They can also be used to harness the immune system to fight cancers. TALENs are used to generate T cells resistant to chemotherapeutic drugs that also show anti-tumor activity (AAS, 2017).

Theoretically, TALENs can correct errors at individual genetic loci via HDR using an exogenous template (Bloom et al., 2015). However, in reality, TALENs are limited by a lack of efficient delivery mechanisms, unknown immunogenic factors, and unspecific binding (Congressional Research Service, 2018).

TALENs can also be combined with genome editing tools such as meganucleases. The DNA-binding region belonging to a TAL effector can be combined with the cleavage domain of a meganuclease. This creates a hybrid architecture that combines the ease of engineering and specific DNA-binding activity of a TAL effector with the low site frequency and specificity of a meganuclease (AAS, 2017). TALENs can also be used to understand the biology and function of organisms and their systems.

TALENs were used to modify the *Gryllus bimaculatus* genome (Watanabe et al., 2017). TALENs were also injected into pre-blastoderm embryos of the mosquito *Aedes aegypti* to target an eye pigmentation gene (Aryan et al., 2013a, 2013b). This reverse genetic experiment resulted in between 20 to 40% of survivors having white eyes. TALENs were also used to engineer UGT1A1-deficient mouse liver cell lines for the study of the CN1 disease (Porro et al., 2014). They also achieved complete silencing of diacylglycerol acyltransferase-1 (DGAT1) to abrogate the entry of HCV in Huh-7.5 cells (Chang et al., 2016).

TALENs have also been used in crop improvement. In rice, TALENs were used to reduce the effects of *Xanthomonas oryzae*, which is a pathogen responsible for blight disease. In barley,

TALENs were used to target the promoter region of a phytase gene HvPaphya and used to induce site-directed mutagenesis in soybean (Pratap and Sharma, 2016).

Advantages of TALENs

TALENs are precise, readily available, and easy to design. They have high-affinity rates of around 96% (Rode et al., 2019). TALENs take a short time to design (about two days) and can be produced in large numbers, ranging in hundreds at a time. This makes the creation of TALENs libraries easy (Bloom et al., 2015).

TALE-based DNA-binding modules are more efficient than other gene-editing tools because they are specific and exhibit low off-target activity. TALENS are less cytotoxic and have a wide targeting range including small DNA sequences such as DNA sequences coding microRNA and enhancers, which may lack in targetable sites for ZFNs or CRISPR-Cas (Bartlett and Root, 2015).

Disadvantages of TALENs

TALENs are large, making them difficult to deliver to many different cell types (AAS, 2017). The cDNA encoding a TALEN is around 3 kb in size (Carter, 2016). This makes them less attractive for therapeutic applications in viral vectors such as AAV that have a cargo limit of 5 kb (Liddicoat, 2016). The repetitive nature of TALENs impairs their ability to be packaged and delivered using viral vectors. Diversifying the TALE repeats coding sequences helps overcome this hurdle (Kruminis-Kaszkiel et al., 2018).

TALENs are available, but they need experience to make the final TALENs efficient. Purchasing them is expensive, although they can be made in-house (Prowse et al., 2017; Congressional Research Service, 2018). However, TALENs can produce off-target mutations (Goold et al., 2018; Congressional Research Service, 2018; Pu et al., 2015).

Due to undesired off-targets, various computational algorithms have been developed for identifying potential off-target sites spanning a wide range of approaches and techniques. The search parameters can be fixed by the user to work with any TAL effector or TAL effector nuclease architecture. These algorithms include TALE-NT, idTALE, E-TALEN, and TALENgetter. To reduce off-targets, approaches like FokI variants and nickases have been developed. The assembly cost of TALENs is very high and laborious making them unavailable to many. Additionally, multiplex editing of TALENs is very difficult (Gaj et al., 2013).

4
MegaTALs

Introduction and History

Meganucleases were discovered in the late 1980s. They are enzymes in the endonuclease family. Meganucleases are characterized by their capacity to recognize and cut large DNA sequences ranging from 14 to 40 base pairs, making them very specific. They are found more commonly in microbial species. However, it is rare to find the exact meganuclease that is required to act on a chosen target DNA sequence. Mutagenesis and high throughput screening methods can be used to create meganuclease variants with the ability to recognize unique sequences. Various meganucleases can be fused to create hybrid enzymes that recognize new sequences. A method known as rationally designed meganuclease can also be used to alter DNA-interacting amino acids of the meganuclease to design sequence-specific meganucleases. Computer models can be used to accurately predict the activity of modified meganucleases as well as the specificity of the recognized nucleic sequence (Odongo, 2019).

MegaTALs comprise a novel rare cutting endonuclease platform that couples a TALE DNA-binding domain with a homing endonuclease (HE) cleavage domain (Guha et al., 2017). They induce highly specific gene modifications. Incorporation of the DNA-binding domain from TAL effectors into hybrid nucleases producing "megaTALs" results in a combination of the ease of engineering and high DNA-binding specificity of a TAL effector with the high cleavage efficiency of meganucleases. In addition, meganucleases have been fused to DNA end-processing enzymes to promote error-prone NHEJ and to increase the frequency of mutagenic events at a given locus (Takeuchi et al., 2014). Meganucleases are endodeoxyribonucleases naturally found in two forms, dimers or single-chain monomers, with five families of meganucleases, the family LAGLIDADG being the best characterized and largest (Porteus, 2019). These owe their name to a conserved amino acid sequence. It contains the most specific cutters, which include *I-SceI* of yeast *S. cerevisiae*.

MegaTALs typically have an 8–12 repeat TAL array appended to the N terminus, making them ~2.5 kb per monomer. The relatively small size and monomeric structure of meganuclease-based platforms make them attractive from a delivery standpoint and afford them the ability to be packaged into a variety of gene transfer vectors (Townson, 2017). Meganucleases have a large recognition site of double-stranded DNA sequences of 12 to 40 base pairs, and their recognition sequences generally occur once in any genome, making them rare cutters. An example is the 18-base pair sequence that is recognized by the I-SceI meganuclease, which would require a genome that is about 20 times the size of the human genome for it to be found once by chance. However, sequences that have a single mismatch can occur about three times per human-sized genome and meganucleases are thus classified as the first class of nucleases that are sequence specific and which create targeted DSBs in eukaryotic genomes (Daboussi et al., 2015).

Ease of Use

Meganucleases can be used to target any plant or animal genome, although they have not been used extensively in many organisms, including goats and sheep (Bevacqua et al., 2013; Wang et al., 2014). The low usage of meganucleases in genome engineering can be attributed to complexities in engineering them and protein redesign to direct them to new or novel DNA sequences (Petersen, 2017). In practice, meganucleases are difficult to engineer because the DNA-binding and endonuclease activities reside in the same domain, and their development has stalled compared to other programmable nucleases (Townson, 2017). Less complicated genome editing tools that include TALENS, ZFNs, and CRISPR are at times preferred over meganucleases. Strategies have been developed to engineer meganucleases with new properties and DNA-binding specificities to widen their applications (Galetto et al., 2009).

DOI: 10.1201/9781003165316-4

Efficiency

MegaTALs have been able to achieve efficient gene editing with optimal reagents in cell culture lines where high-level delivery and toxicity are not a concern. In primary human T cells, editing rates have been reported to be greater than 70% for megaTALs when electroporating mRNA (Porteus, 2016).

Specificity

A universal prerequisite for all molecular scissors applied in gene editing is to have a unique target site in the genome to reduce or eliminate sheering the genome into unintended pieces and inducing toxicity (Boglioli and Richard, 2015). The fusion of the site-specific meganuclease cleavage head, with additional affinity and specificity provided by the TAL effector DNA-binding domain, facilitates the transformation of megaTALs into hyperspecific and highly active molecular scissors compatible with multiple cellular delivery methods (Boissel et al., 2014).

Meganucleases, found commonly in microbial species, have the unique property of having very long recognition sequences (>14 bp), thus making them naturally very specific. The integrated nature of megaTALs' DNA recognition and cleavage makes them potentially highly specific. It also makes them complex to reprogram for the recognition of novel DNA targets, and they have lower on-target cleavage efficiency than other genome targeting endonucleases (Guha and Edgell, 2017).

Versions of Meganucleases

A single-chain modular nuclease architecture termed "megaTAL" was designed in which the DNA-binding region of a TAL effector is appended to a site-specific MN for cleaving a desired genomic target site (Guha et al., 2017). The latter synthetic version of an MN provides a modular design, separating the endonuclease and DNA-binding activities. The "MegaTev" is a recent architecture that has been generated and involves the fusion of the DNA-binding and cutting domain from a meganuclease (Mega, I-OnuI) with another nuclease domain derived from the GIY-YIG HEase (Tev, I-TevI). This protein was designed to position the two cutting domains ~30 bp apart on the DNA substrate and generate two DSBs with non-compatible single-stranded overhangs for more efficient gene disruption. More recently the TevCas9, similar to the MegaTev concept, was designed by Wolfs in 2017, in which the Tev endonuclease domain is attached to the Cas9 nuclease This hybrid nuclease when introduced within human embryonic kidney cells (HEK293) along with appropriate guide RNAs, has been shown to delete 33 to 36 bp of the target site, thereby creating two noncompatible DNA breaks at moderately higher frequencies (40%) and promises to favor genome editing events by avoiding the creation of compatible "sticky" ends, which lead to a failed attempt of genome editing domain (Friedrichs et al., 2019; Wolfs, 2017).

Construct

The megaTALs construct includes using a repeat variable diresidue (RVD) plasmid library and destination vector and protein linkers. Plasmids can be modified to allow the assembly of TAL effectors that target each specific nucleotide (Porteus, 2019; Guha et al., 2017).

Principle Behind MegaTALs

MegaTALs combine the DNA-binding domains from TALENs with the high cleavage efficiency of meganucleases in a single chimeric protein. This creates large variants such as LHE, I-SceI, DmoCre, and E-DreI. These can provide site-specific nucleotide cleavage (Khan, 2019; Aglawe et al., 2018; Wolfs, 2017). DNA-binding domains from TALENS are easy to design. A TAL effector is fused with the truncated N- and C-terminal domains through a short linker to the N-terminus of an mn (I-AniI). The N-terminus of an mn displays a KD of approximately 90 nM for its cognate target site (Boissel et al., 2014). They have greater flexibility compared to native MgN. MegaTALs have a small size, specificity, and high cleavage efficiency, making them preferable for therapeutic applications (Wang et al., 2014). Their small size (containing both catalytic and DNA-binding capabilities) makes them compatible with most of the available cell delivery strategies. MegaTals also use just one protein to achieve efficient gene editing. However, they are not frequently used in research applications due to difficulties in their design.

This is attributed to the fact that the creation of new meganucleases requires advanced protein engineering. There is a need to alter specificities of the naturally occurring meganucleases. This is done through a series of mutations whose outcome is difficult to predict. Altering their specificities can also be achieved by creating chimeric proteins that combine active sites through domain swapping between two subdomains of different HEs. Typically there are two naturally occurring meganucleases. MegaTALs' additional TALE RVD recognition modules can improve their effective specificity. This is because their unique 39 overhangs, which are generated by a meganuclease head (including by megaTALs), can enhance HDR rate relative to FokI or Cas9 cleavage. Meganuclease engineering can be improved by providing a programmable second DNA-targeting domain (Wolfs, 2017; Guha and Edgell, 2017; Hoban and Bauer, 2016; Boissel et al., 2014).

Applications of MegaTALs

MegaTals can be used as a genome editing tool, including the correction of variants that cause monogenic diseases, the enhancement of chimeric antigen receptor (CAR) T-cell therapy, and cell-based regenerative medicine. They can be used in gene targeting and gene correction of diseases (Georgiadis and Qasim, 2017; Li et al., 2020).

MegaTALs have a therapeutic application in preventing entry of HIV into host CD4 cells through gene editing at the CCR5 locus of primary human T cells. A CCR5-specific megaTAL and recombinant adeno-associated virus (rAAV) were used to deliver the donor template. The resulting protocol allowed efficient target of a range of expression constructs to the CCR5 locus in primary T cells and adult mobilized CD34+ peripheral blood stem cells (PBSCs). Results demonstrated that high-efficiency targeted integration is feasible in primary human hematopoietic cells and highlights the potential of gene editing to engineer T cell products with myriad functional properties. They also exhibit the potential to use gene-modified cells to reconstitute a patient's immune system and provide protection from HIV infection (Sather et al., 2015).

MegaTALs have also been used for research on adoptive immunotherapy strategies, which are based on the retargeting of autologous T cells to recognize tumor antigens. Varying degrees of gene disruption were observed among reagents generated from T cell Receptor antigen-targeted TALEN, T cell receptor antigen-targeted megaTAL, and Tcell receptor antigen-targeted CRISPR-Cas9, with a paucity of off-target effects. The approach was further validated by generating T cell receptor null cells that expressed a CD19 chimeric antigen receptor (CAR) construct and a translationally compatible manufacturing process to demonstrate the capacity to generate and expand large numbers of engineered cells. The megaTAL and CRISPR-Cas9 reagents showed the highest disruption efficiency integrated with low levels of toxicity and off-target cleavage, and they were used for a translatable manufacturing process to produce safe cellular substrates for next-generation immunotherapies (Osborn et al., 2016).

The combined use of engineered antigen-targeting moieties and innovative genome editing technologies have recently shown success in a small number of clinical trials targeting HIV, hematological malignancies, and renal and ovarian cancers. They are now being incorporated into existing strategies for other immunotherapies (Delhove and Qasim, 2017).

Advantages and Disadvantages

Meganucleases have long recognition sequences, thus rendering them very specific, and this can be exploited to make site-specific DNA DSB in genome editing. However, meganucleases are insufficient to cover all possible target sequences (Bahrami and Nafaji, 2019; Li et al., 2020). MegaTALs have the benefit of causing less toxicity in cells than methods such as ZFN because of more stringent DNA sequence recognition. However, the construction of sequence-specific enzymes for all possible sequences is costly and time consuming. Meganucleases cause less toxicity in cells than ZFNs. Other genome editing modalities, including TALENs, require two proteins to carry out DNA cleavage, while megaTALs only need a single molecule. One essential drawback for this class of enzyme is its non-modular configuration, since the DNA recognition and cleavage functions can be, in part, intertwined in a single protein domain (Guha et al., 2017; Porteus, 2016, 2019).

5
CRISPR

History

A clustered regularly interspaced short palindromic repeats (CRISPR) system was discovered in 1987 as a defense system in *E. coli*. It acts against viruses. The CRISPR system is found only in bacteria and archaea but not in eukaryotes and viruses (Jansen et al., 2002). The variations of CRISPR-associated proteins include Cas3, Cas6, Cas8, and Cas9 (Kruminis-Kaszikiel et al., Mudziwapasi et al., 2018; Sorek et al., 2013; Kunin et al., 2007). CRISPR Cas9 is based on a bacterial Cas9 from *Streptococcus pyogenes*. Cas stands for CRISPR associated system.

CRISPR loci are short sequences of the viral or plasmid genomes that are integrated into the CRISPR locus of the bacterial or archaeal genome. They were first identified by scientists working in the fermentation industry, where prokaryotes are very essential to the production of fermented products (Kunin et al., 2007). Through comparative genomic analysis of different *S. thermophilus* strains (a microbe used in producing yogurt), scientists identified a highly variable locus in the genome of these bacteria (Sorek et al., 2013, Kunin et al., 2007; Zhang et al., 2013). This highly variable region has two distinct features that include many noncontiguous repeats that are separated by variable sequences and are known as spacers. The researchers then found out that the spacer sequences matched those found in bacteriophage (viruses that infect bacteria) genomes (Mohanty et al., 2019). When the researchers compared phage resistant and phage sensitive *S. thermophilus*, the phage resistant bacteria had spacer sequences that matched regions of that phage's genome (Horvath and Barrangou, 2010). These repeats were discovered in 1980 and later confirmed in 2007 by Barrangou and colleagues. They demonstrated that *S. thermophiles* could acquire resistance against a bacteriophage by integrating a genome fragment of an infectious virus into its CRISPR locus (Simon et al., 2018).

In natural prokaryotic systems, the CRISPR system is used by host bacteria to remove the viral DNA of invading viruses. The bacteria take pieces of DNA from invading viruses and use it to create DNA segments called CRISPR arrays. The CRISPR arrays enable the bacteria to identify the viruses upon the second attack. Once the viruses attack again, the bacteria produce RNA segments from the CRISPR arrays to target the viruses' DNA. The bacteria will use Cas9 or similar enzymes to cut the DNA apart, destroying the virus. It degrades foreign nucleic acids entering the cell, thereby preventing bacteriophage infection, conjugation, and natural transformation (Sampson et al., 2014; Cong et al., 2013; Zhang et al., 2013). It targets repeats associated with viral insertions as a means of the immune system to combat infection.

There are many protein-coding gene clusters near CRISPR arrays. These genes are highly conserved among bacteria and archaea (Jansen et al., 2002). The function of these loci was uncovered in 2005. The unique spacer regions that are found in CRISPR arrays were found to be mapped to phage genomes (Bolotin et al., 2005; Pourcel et al., 2005). This proved that CRISPR is an adaptive immune response to phage infection that works through an RNA-guided process. The CRISPR arrays are regularly interspaced, while the spacers lack sequence conservation. The arrays are variable in number. They differ in the number of repeats (Pearson et al., 2015; Marraffini and Sontheimer, 2008).

The molecular mechanism of this immune response was proposed in 2012. It was demonstrated that CRISPR arrays are transcribed into RNA. The RNA is then cleaved and loaded into Cas proteins (Cas9, in this case). The RNAs are noncoding, and they guide the Cas nuclease to induce site-specific DNA cleavage. The Cas protein cuts DNA, making a DSB. This system can be reprogrammed by redesigning the RNA to target novel sequences. These RNAs are called single guide RNA (sgRNA) (Jinek et al., 2012). RNA-guided mutation in eukaryotic cells was first demonstrated in 2013 (Cong et al., 2013). The events that occurred during the discovery of CRISPR are summarized in Figure 5.1

The CRISPR system was initially used for typing purposes. This is because it is a low cost and a high-resolution approach. Bacterial and archaea strains can be differentiated based on the

DOI: 10.1201/9781003165316-5

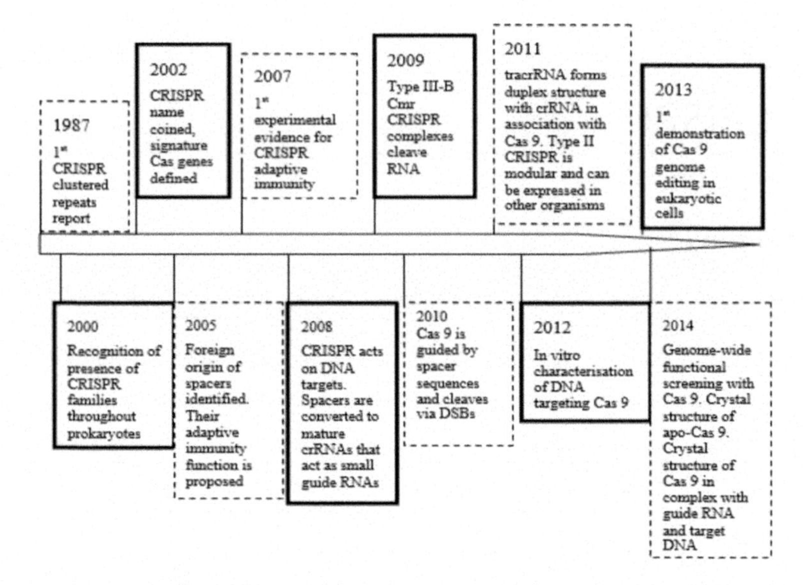

Figure 5.1 History of CRISPR.

absence or presence of CRISPR arrays, thereby improving disease diagnosis (Shariat et al., 2013; Horvath and Barrangou, 2012). Now CRISPR has been adapted for genome editing. It has two predecessors, namely ZFNs and TALEs. Several patents have been applied for and granted beginning in 2012 in various jurisdictions for various innovations related to the CRISPR system.

Versions of CRISPR

CRISPR systems can be classified into class 1 and class 2. Class 1 has three types, which are type I, type III, and type IV, while class 2 has type II and type V. Class 1 system are characterized by their use of a multi-subunit crRNA effector complex while on the other hand, the class 2 systems use a single subunit effector molecule (Haurwitz et al., 2010).

There are three types of CRISPR based on gene sequence expression for Cas protein with type I for Cas3 and Cas6, type II for Cas9, and type III for Cas10. The type II CRISPR nuclease has three components. These are Cas9, crRNA, and trRNA. This makes it useful for genome editing (Mohanty et al., 2019; Jiang and Doudna, 2017; Harvath and Harvarth, 2010). Cas9 is a ribonucleoprotein (RNP) that binds and cleaves DNA at sequences that bind to the crRNA of the Cas9 RNP. It is an RNA-guided endonuclease that is guided by guide RNAs (gRNA) (Mohanty et al., 2019; Doudna and Jiang, 2017; Doudna and Charpentier, 2014).

Three variants of the Cas nucleases have been adopted for use in genome editing. These include the wild-type Cas9. It can site-specifically cleave double-stranded DNA. Cas9 searches the invading DNA for sequences complementary to the crRNA and then binds to sequences in the invading viral or plasmid genome known as protospacer adjacent motifs or PAMs (Frank and Charpentier, 2016; Crowther et al., 2020; Doudna and Jiang, 2017). Different Cas9 proteins from different species of bacteria or archaea recognize different PAM sites. *S. pyogenes* Cas9 (SpCas9) recognizes a 50-NGG-30 PAM. It is the most commonly used for genome editing processes (Mohanty et al., 2019). Two critical arginine residues in SpCas9, Arg1333, and Arg1335, interact with the guanine nucleobases of the PAM on the noncomplementary strand (Rodriguez, 2017; Crowther et al., 2020). This interaction between the guanines of the PAM and the arginines in SpCas9 positions the phosphate of the DNA backbone 5' to the PAM to interact with a phosphate-lock loop in Cas9 and facilitate DNA strand unwinding. If the DNA is complementary to the guide RNA, an RNA:DNA hybrid forms, called an R loop, and cleavage then follows (Lander, 2016; Gherardin, 2017; Horvath and Barrangou, 2010).

DNA cleavage then occurs, and it results from the action of two different Cas9 nuclease domains. The HNH domain nicks the DNA strand that is complementary to the crRNA, and the RuvC-like domain nicks the strand that is not complementary to the crRNA. Cas9 cleaves the DNA 3 base pairs upstream of the PAM, resulting in a blunt-end cleavage of DNA. The cleaving of the DNA is deleterious to the invading plasmid or virus, and this results in the degradation of and subsequent protection against these invaders (Crowther et al., 2020).

When host DNA is cut, the cell repair machinery is activated. The DNA repair will proceed via the NHEJ pathway or the HDR (Gherardin, 2017). NHEJ can occur through canonical NHEJ (C-NHEJ), which ligates or essentially glues the broken ends back together. NHEJ can result in indels, which can cause mutations (Martinez-Lage et al., 2018; Rodriguez, 2017). Additionally, there is an alternative nonhomologous end joining pathway (alt-NHEJ), in which one strand of the DNA on either side of the break is resected to repair the lesion. Both of these repair methods are error-prone, meaning that ultimately the lesion is repaired with potential for errors (Horvath and Barrangou, 2012; Makarova et al., 2011).

Alternatively, the HDR repair pathway is used. The HDR pathway requires that a piece of DNA with ends that are homologous to the ends on the cut position with the DSB be available (Mohanty et al., 2019; Frank and Charpentier, 2016). The homologous DNA can be used as a template to repair the breakthrough in the HDR pathway (Martinez-Lage et al., 2018; Rodriguez, 2017; Crowther et al., 2020). Usually, this repair mechanism happens after DNA replication but before cell division, so the break can be repaired by the newly replicated sister chromatid without any mutations occurring. This form of repair can, however, then be exploited to introduce precise edits or large indels by introducing a donor template for repair (Lemay et al., 2017; Malnoy et al., 2016; Woo et al., 2015, Gherardin, 2017). Therefore, by making a cut at a specific locus and taking advantage of the cellular DNA repair pathways, there is the potential to generate targeted mutations and insert sequences of interest imperfectly, resulting in indels (Deborah et al., 2018; Crowther et al., 2020). Pathways for genome editing using CRISPR-Cas9 are shown in Figure 5.2.

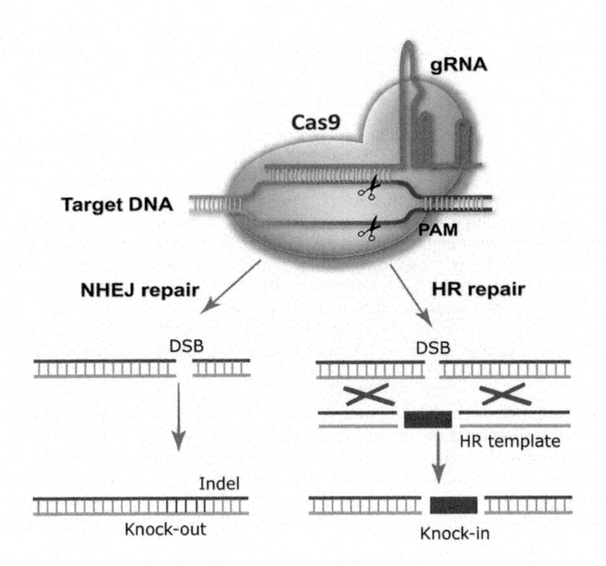

Figure 5.2 Pathways for genome editing using CRISPR-Cas9. (From Ding et al., 2016.)

Cas9-induced DSB can be repaired by both the NHEJ and the HDR. Figure 5.2 shows the sequence of a targeted genomic locus in relation to the PAM (50-NGG-30) site. It also shows a cartoon representation of a crRNA, tracrRNA, and Cas9 protein assembly. Figure 5.2 additionally shows NHEJ resulting in random indels. It also shows HDR resulting in precise researcher-designed edits. To achieve HDR, there is a need to also introduce a repair template that contains the desired edit that the HDR machinery of the cell uses to repair the induced (Mohanty et al., 2019; Frank and Charpentier, 2016). The second type is Cas9D10A. It is a mutant form and has nickase activity. Thus it only cleaves one strand of double-stranded DNA and does not activate NHEJ. It was developed by Cong and colleagues (2013; Peng et al., 2018; Lander, 2016). The third type is the dCas9. It is a nuclease-deficient Cas enzyme; thus, it cannot cut the targeted DNA. Mutations H840A in the HNH domain and D10A in the RuvC domain inactivate cleavage activity but do not prevent DNA binding. It is used to sequence-specifically target regions of the genome without cutting them. Thus it can be used to visualize DNA and to activate or silence genes (Martinez-Lage et al., 2018; Rodriguez, 2017).

Cas proteins can be classified into three types. These are types I, II, and III. Type I and type III systems use Cas6 to break the long strand of crRNA into multiple smaller crRNAs (CRISPR RNA) during the production of the crRNA. Type II systems use the RNase III cutting mechanism. This system requires an additional small RNA, tracrRNA (trans-activating CRISPR RNA), which is complementary to the repeat sequence. Matured crRNA associate with Cas proteins. Each system, therefore, utilizes a different mechanism to generate crRNA and Cas proteins that catalyze the nucleic acid cleavage (Makarova et al., 2011).

The Class 1 system represents about 90% of CRISPR-Cas loci and is more widely present than Class II in both bacteria and archaea (He et al., 2020; Cornell, 2020). Type I is most widespread within the Class I system and functions as a CRISPR RNA (crRNA)-bound multiprotein complex, termed Cas complex for antiviral defense (Cascade), and as a Cas3 endonuclease, which is recruited upon target binding by Cascade to cleave foreign DNA (Murugan et al., 2017; Nakade et al., 2017).

Cas3 is an ATP-dependent single-strand DNA (ssDNA) translocase/helicase enzyme that degrades DNA as part of CRISPR based immunity (Koonin et al., 2019; Fu et al., 2017). It is a "signature" protein of class 1 CRISPR systems and functions in a complex known as Cascade with other Cas genes and a targeting RNA to degrade viral DNA. CRISPR-Cas3 can efficiently delete long swaths of DNA from a targeted site in the human genome (Loeff et al., 2018; Rath et al., 2018). This ability is superior to that achieved with the more common CRISPR-Cas9 systems. Cas3 possesses helicase and nuclease activity, predominantly triggered by several thousand base pair deletions upstream of the 5′-ARG protospacer adjacent motif (PAM), without prominent off-target activity (Xiao et al., 2018).

The Cas3-mediated directional and broad DNA degradation can be used to introduce functional gene knockouts and knock-ins (Ifuku et al., 2018). An example of potential therapeutic applications is Cas3-mediated exon-skipping of the Duchenne muscular dystrophy (DMD) gene in patient induced pluripotent stem cells (iPSCs) (Kosicki et al., 2018). Cas3 may serve as a unique genome editing tool in eukaryotic cells distinct from the Class 2 CRISPR system. *E. coli* Cas3, with just a single crRNA, can induce large deletions more efficiently than Cas9. Cas3 achieves more efficient genome editing than Cas9 from a distance of a few dozen to a few hundred bp upstream of the targeted site, while Cas9 can enforce more efficient genome editing at the target site (Morisaka et al., 2019). Thus, Cas3-mediated genome editing may be useful when targeting specific sequences that are difficult-to-design on-site sgRNAs, such as sites far from PAM sequences, repetitive sequences, or transposon elements.

Type 1 systems are normally found in bacteria and archaea, and the gene that defines the system is the Cas3 gene; its protein exhibits helicase activity in the presence of ATP, and it degrades single-stranded DNA. In type I, Cas proteins cleave the pre-crRNA into mature crRNA. The Cas3 protein is the nuclease unity of the effector complex, and it contains a histidine aspartate region with a metal ion binding site. Type I systems are divided into seven subtypes ranging from I-A to I-F and I-U, with I-U being the uncharacterized subtype because of unknown systems.

Type I systems associate with many Cas proteins, forming a complex that recognizes foreign DNA, triggering its degradation. Type II is the most commonly used system for engineering purposes, and they are defined by the Cas9 genes. These systems have been further classified into three subtypes, which are type II-A systems, type II-B systems, and type II-C systems. Among these systems, type II-A and type II-C are the most abundant and common. Type II-A systems are characterized by the csn 2 gene, while type II-B systems have a Cas4 gene in place of the csn 2 gene. Subtype II-C has a three protein-coding gene. Type II systems process the pre crRNA through a trans-encoded crRNA that is complementary to the repeat sequence of the pre-crRNA, leading to its cleavage by dsRNA-specific ribonuclease RNase III in the presence of Cas9 (Deltcheva et al., 2011).

Type II systems rely on the association of both crRNA and tracrRNA with Cas9 as a single multifunctional protein for them to recognize and degrade foreign DNA. Type III is defined by the Cas10 gene; its protein forms the largest subunit of the effector complex, and it contains the HD nuclease domain. These type III systems have been subdivided into type III-A, type III-B, type III-C, and type III-D. Both type III-A and type III-B target RNA and DNA. Type III systems require association with many Cas proteins. However, type III systems target foreign RNA, not foreign DNA. This enables the bacteria and archaea to target RNA viruses. Type IV systems are found in some bacteria, like *A. ferroxidans*. The type V systems contain the cpfl gene. This gene is homologous to Cas9 and has a RuvC-like nuclease domain, though it lacks the HNH domain.

Mechanism of the CRISPR System

The CRISPR mechanism has three stages. The first stage is the acquisition stage (adaptation), where a distinct DNA from invading viruses (foreign DNA) is incorporated into the bacterial genome at the CRISPR loci on the leaders' side (Barrangou et al., 2007). The integrated sequences will function as a genetic memory and will help the host quickly fight off a future infection by the same virus with the same sequence. DNA next to PAM sequence (3 prime of the crRNA complementary sequence) which is 2–5 nucleotides long is cut out of the invading virus DNA (Mali et al., 2013; Mojica et al., 2009). It is then incorporated into a CRISPR locus amidst a series of short repeats, approximately 20 base pairs (Jinek et al., 2012). The second stage is the crRNA biogenesis (or expression). The CRISPR loci are transcribed and processed into crRNA (Marraffini, 2015; Makarova et al., 2011). In the leading sequence flanking the CRISPR locus, there is a promoter sequence. It synthesizes a noncoding primary transcript of the CRISPR locus (Brouns et al., 2008; Pul et al., 2010). The transcript is called a pre-CRISPR RNA (pre-crRNA) (Deltcheva et al., 2011). In turn, tracrRNA that is complementary to the pre-crRNA and that pairs with its repeat sequence cleaves the pre-crRNA into mature crRNAs by the activities of endogenous RNase III in the presence of a Cas protein. The crRNAs guide effector endonucleases to target DNA based on sequence complementarity (Hill and Charpentier, 2016; Shabbir et al., 2016; Rath et al., 2015). The interference stage (maturation) is the final stage. The Cas endonuclease complexed with a crRNA and tracrRNA cleaves foreign DNA containing a 20 nucleotide crRNA complementary sequence that is adjacent to a PAM sequence. The cleaving process starts with the activation of the Cas protein. This occurs through base-pairing between tracrRNA and crRNA in the repeat sequence (Jinek et al., 2012). The DSB in the target DNA triggers the cell's repair mechanism (Koonin and Makarova, 2019; Kuzma et al., 2017). The Cas protein is guided by crRNA to target DNA. In the absence of an adjacent PAM sequence, the Cas protein may not cut the foreign DNA even though they are fully complementary to the crRNA (Makarova et al., 2018; Mali et al., 2013; Mojica et al., 2009).

Cutting of foreign DNA is done by two nuclease domains on the Cas protein. These are RuvC and HNH nuclease domains (Makarova et al., 2006). Mutations in these domains abolish the ability of Cas proteins to interfere with foreign DNA (Sapranauskas et al., 2011). The HNH domain cleaves the complementary strand three base pairs upstream of the PAM sequence of the invading nucleic acid. The RuvC domain cleaves the noncomplementary strand (Jinek et al., 2012).

The CRISPR-Cas system provides immunity to phages in prokaryotes, and its main features can be described by three distinct stages as shown in Figure 5.4.

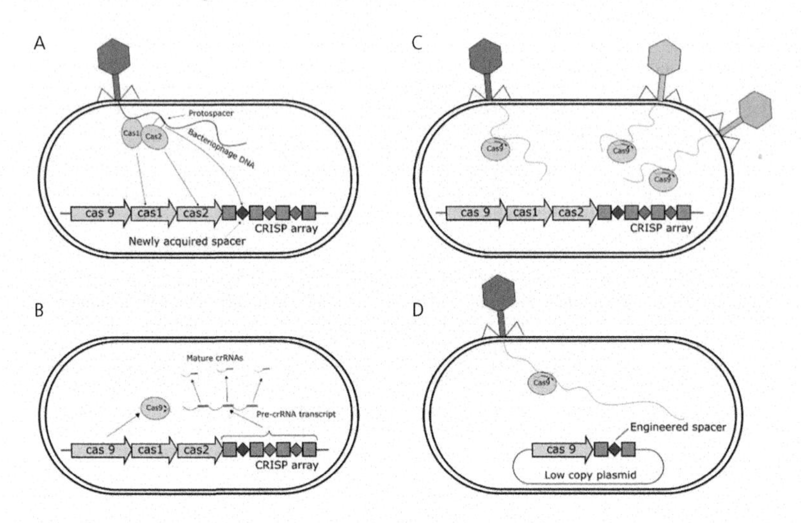

Figure 5.4 Mechanism of the CRISPR system in prokaryotes (Payne et al., 2018)

(A) *Acquisition:* When a cell gets infected by a phage, a protospacer on the invading phage DNA is recognized by Cas1 and Cas2. The protospacer is cleaved out and ligated to the leader end (proximal to the Cas genes) of the CRISPR array as a newly acquired spacer (diamond).

(B) *Processing:* The CRISPR array is transcribed as a pre-crRNA and processed by Cas9 (assisted by RNaseIII and trans-activating RNA) into mature crRNAs.

(C) *Interference:* Mature crRNAs associate with Cas9 proteins to form interference complexes that are guided by sequence complementarity between the crRNAs and protospacers to cleave the invading DNA of phages whose protospacers have been previously incorporated into the CRISPR array.

(D) *A truncated version of the CRISPR system on a low copy plasmid may lack Cas1 and Cas2 genes:* It can be engineered to target a protospacer on the T7 phage chromosome to provide bacteria such as *Escherichia coli* cells with immunity to the phage. The susceptible strain will contain the same plasmid, except the spacer will not target the T7 phage chromosome (Makarova et al., 2011).

Genome Editing Using CRISPR

The requirements for CRISPR-Cas genome editing are: a Cas nuclease, gRNA, and template for HDR-based repair. In practice, active Cas nuclease complexes can be generated in the cell.

The first step to generating the desired mutation is designing the gRNA, and there are numerous guidelines to consider when creating a gRNA. Most importantly, the 20-nucleotide target region of the gRNA must be adjacent to a PAM site, 50-NGG-30 in the case of SpCas9 (Doech et al., 2016; Haeussler et al., 2016; Keller, 2017). Therefore, one must identify the genomic region where the desired mutation is to be generated and select a 20-nucleotide target in that region that is adjacent to a PAM site (Doech et al., 2016). If target sequences are not specific enough, Cas9 can bind and cut in a different place than intended, and this may result in background mutations that could confound experimental results. Once optimal gRNAs have been designed, Cas9 and sgRNAs can be introduced using three different strategies: the sgRNA or crRNA and tracrRNA and Cas9 can be expressed as DNA, RNA, or RNA/protein complexes with the nucleic acid and/or protein introduced using a variety of methods. It can be introduced through microinjection in worms, fruit flies, and zebrafish. Alternatively, it can be introduced through electroporation or transfection in mammalian cell culture (Haeussler et al., 2016; Keller, 2017). This is illustrated in Figure 5.5.

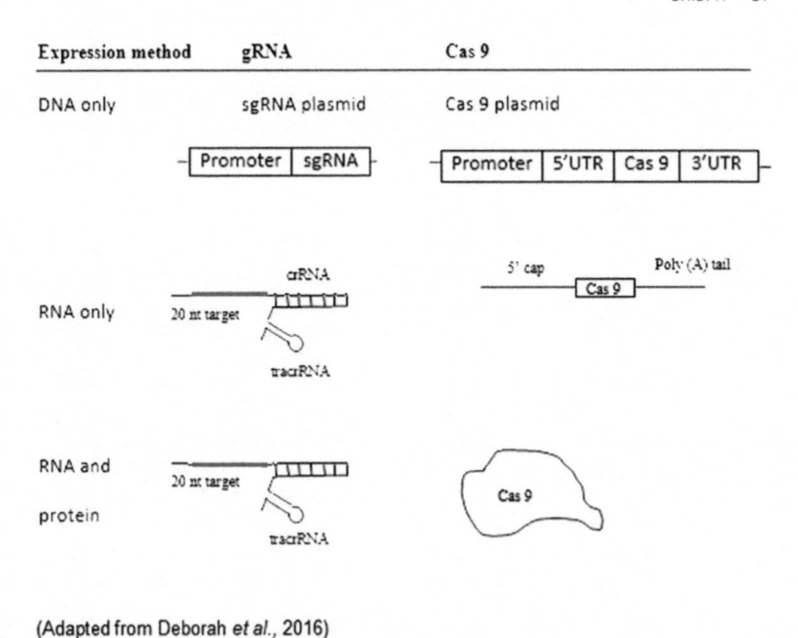

Figure 5.5 Methods of introducing CRISPR components into cells.

Different methods can be used to introduce CRISPR-Cas9 components. CRISPR guides and the Cas9 protein required for genome editing can be introduced into organisms or cells both as DNA plasmids, both as RNA molecules, or RNA and protein complexes (RNPs).

For all the methods described, a single-stranded or double-stranded DNA can be included as a homology-directed repair template to generate a researcher-designed edit. Cas9 has two nuclease domains, each cutting a different strand of DNA (Doech et al., 2016). The wild-type Cas9 contains two nuclease domains, RuvC and HNH, which each cut a different strand of the DNA. When the RuvC nuclease domain is mutated, Cas9 will act as a nickase and produce a nicked DNA product (Mali et al., 2013; Mojica et al., 2009).

The binding of the crRNA to the DNA by base pairing of a 20 nucleotide region at the 5' end brings about the specificity of the Cas9 system. This is followed by rearrangement of the nuclease lobe of Cas9 (containing the HNH and RuvC domains), leading to a DSB of the target DNA. However, the target site must be adjacent to a PAM. The PAM consists of random nucleotides and two guanines (NGG) (Mali et al., 2013; Mojica et al., 2009). Thus, the tracrRNA molecule and crRNA function as a scaffold onto which the Cas9 binds. After site-specific DNA cleavage, NHEJ and HDR result in indels. Inhibition of expression of the targeted gene can result from the introduction of indels through the NHEJ and the resulting frameshift. The intervening genomic DNA is deleted through the introduction of two DSBs, resulting in activation or deactivation of gene function. HDR allows for highly specific insertion of large or small DNA sequences at a predetermined locus after generation of a DSB. These come from a corresponding repair template. Reduced off-target effects have been observed from engineering of the Cas9 nuclease to a nickase variant. This is done using a paired nicking strategy, where gene excision requires recognition of two different gRNAs on each strand of the target DNA sequence to be successful. dCas9 is a catalytically dead variant developed as a sequence-specific transcriptional repressor and a genomic anchor. It is used for site-specific transcriptional regulation. This occurs when it is coupled with different kinds of genomic effectors. It also aides more precise DNA cleavage when it is coupled with other nucleases (Zhou and Deiters, 2016). This process is illustrated in Figure 5.5.

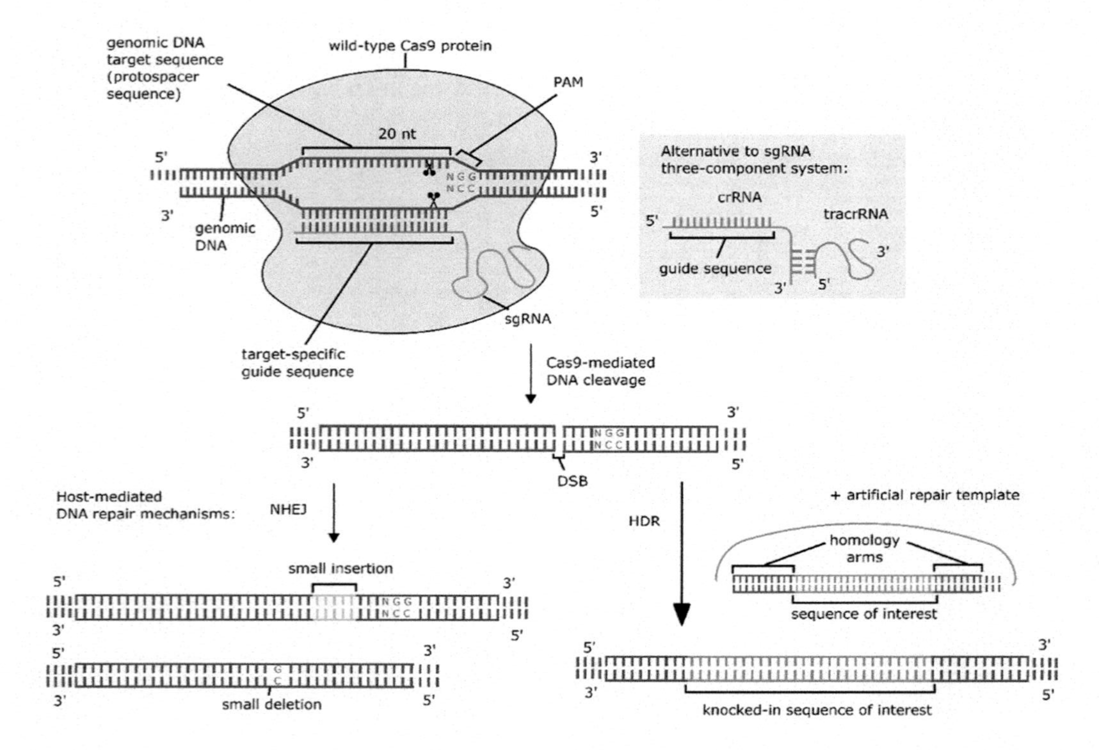

Figure 5.6 A gRNA-directed double-strand DNA cleavage by Cas9 nuclease. PAM: Protospacer adjacent motif, a Cas9 recognition motif (5'-NGG-3') is required at the 3' end of the target DNA sequence; HNH and RuvC: Nuclease domains. Gene mutation

Applications of CRISPR

CRISPR technology can be used to target prokaryotic cells and eukaryotic cells. Thus CRISPR can be utilized for gene editing in multiple organisms such as bacteria, yeast, plants, animals, and human cell lines (Lemay et al., 2017; Malnoy et al., 2016; Woo et al., 2015; Cho et al., 2013). CRISPR technology systems can be used to investigate gene function, gene development, and the development and the pathology of disease (Shao et al., 2016). The gene drive system can also be used to edit specific tissues such as the brain and liver tissues. CRISPR technology can be used to generate simultaneous multiple gene mutations (Li et al., 2013). CRISPR is gaining wide acceptance in CHO cell line engineering, synthetic biology, and genome engineering (Pennisi, 2013). It has many laboratory applications, including the rapid generation of cellular and animal models, functional genomic screens, and live imaging of the cellular genome (Gherardin, 2016; Malnoy et al., 2016; Lemay et al., 2017).

This technology has a great potential to be used to cure animal diseases by disrupting endogenous disease-causing genes, correcting disease-causing mutations, or inserting new genes with protective functions (Rozov et al., 2019; Ray and Felice, 2015). CRISPR-Cas9 has the potential to treat genetic disorders caused by single-gene mutations. Examples of such diseases include cystic fibrosis (CF), DMD, and hemoglobinopathies (Mudziwapasi et al., 2018; Gherardin, 2016). CRISPR technology also has the potential to treat and cure diseases by editing the DNA associated with a particular disease before a baby is born (Foss et al., 2018; Pennisi, 2013; Rozov et al., 2019). CRISPR-Cas9 can therefore be used to edit the genome of preimplantation embryos and to introduce precise genetic modifications into the human germline, including early embryos, sperm, and oocytes, used in in vitro experimental setups (Wang and Yang, 2019; Ray and Felice, 2015; John et al., 2013). These improved technologies may provide solutions in the future for genetic diseases but only when consensus has been met and a regulatory framework has been put in place for treating specific medical implications (Egli et al., 2018; Yusorf, 2018). There also is interest in using CRISPR-Cas9-mediated genome editing in pluripotent stem cells or primary somatic stem cells to treat diseases such as cancer. Research suggests that in the future such an approach could provide a source of cells for bone marrow transplantation to treat β-thalassemia and other similar monogenic diseases (Foss et al., 2018; Pennisi, 2013; Lepore, 2017). CRISPR can also be used to treat polygenic diseases.

Other potential clinical applications of using CRISPR in humans include gene therapy, treating infectious diseases such as HIV, and engineering autologous patient material to treat cancer and other diseases (Gherardin, 2016; Egli et al., 2018; Bruegmann et al., 2019). Although antiretroviral therapy provides an effective treatment for HIV, no cure currently exists due to the permanent integration of the virus into the host genome and its high mutation rates, inter alia. The CRISPR-Cas9 system can be used to target HIV-1 genome activity and inactivate the virus (Lepore, 2017; Nelson et al., 2016). This can inactivate HIV gene expression and replication in a variety of cells, without any toxic effects (Wang and Yang, 2019). The cells can also be immunized against HIV-1 infection. This is a potential therapeutic advance in overcoming the current problem of how to eliminate HIV from infected individuals. It may be possible to use gene therapies or transplantation of genetically altered bone marrow stem cells or inducible pluripotent stem cells to eradicate HIV infection. During the second World Summit of Human Gene Editing, Jiankui He presented the gene-editing project that led to the birth of two baby girls with man-made C-C chemokine receptor type 5 (*CCR5*) mutations. He claimed that he edited the *CCR5* gene to prevent HIV infection in the babies. The father is an HIV carrier, and gene editing in embryos is completely unnecessary to prevent HIV transmission to the fetus (Wang and Yang, 2019). However, there is still debate, and more clarity is needed from these findings (Nelson et al., 2016; Lepore, 2017).

It can be used to cure animal diseases by disrupting endogenous disease-causing genes, correcting disease-causing mutations, or inserting new genes with protective functions. CRISPR can be used to spread synthetic alleles that inhibit female reproduction in unwanted pest populations, invasive species, or disease-carrying organisms. Synthetic alleles that prevent insects from transmitting pathogens can be spread, thereby reducing vector-borne diseases (Yusorf, 2018; Foss et al., 2018; Pennisi, 2013). Editing systems such as CRISPR-Cas9 could carry cargo genes with them to immunize an endangered species against diseases and protect it from particular environmental conditions. It is, therefore, an alternative for protecting endangered species. It can be used to introduce single-point mutations in a particular target gene via a single gRNA. Using a pair of gRNA-directed Cas9 nucleases instead, it is possible to introduce large deletion or genomic rearrangements such as inversions or translocations. It can be used to knock out genes or change the function of genes by adding or replacing sections of DNA. The knockouts can be generated through indels or knock-ins via HDR. The dCas9 version of CRISPR can be used to target protein domains for transcriptional regulation, epigenetic modification, and microscopic visualization of specific genome loci (Bruegmann et al., 2019).

CRISPR technology can also be useful to make genome changes in fertilized animal eggs or embryos (Mali et al., 2013; Kanchiswamy, 2016). This leads to an alteration of the genetic makeup of every cell in an organism and ensures changes are passed on to the subsequent generations (Ray and Felice, 2015; Lepore, 2017; Nelson et al., 2016). CRISPR-based modification drives can introduce a new or desired trait to a target population such as resistance to a certain disease. CRISPR-based suppression drives can control the population by eradicating target populations whose traits are undesirable (Rozov et al., 2019; Ray and Felice, 2015; John et al., 2013). CRISPR gene drives and synthetic biology play a great role in animal science by establishing a line of animals with specific traits based on selective breeding and increasing the prevalence of a genetic variant within a population. This improves animal production. CRISPR can be utilized in the treatment of genetic diseases by correcting the defective genes responsible for these diseases. It can be used to study cancers by turning off the oncogenes and turning on tumor suppressor genes. Exact mutations in different cell lines can be created using CRISPR to model the cancers, thereby assisting in developing effective drugs. In personal medical applications of the CRISPR gene-editing tools, embryonic stem cells can be edited using the CRISPR technique and then reinjected into the patient. In this method, each person is treated according to their genetic makeup, and the faulty genes will be modified directly.

CRISPR technology systems can be used in the epigenome for DNA methylation and histone acetylation. Therefore, the system can also be used to change the epigenome and regulate the expression of specific genes. CRISPR system applications in biodiversity conservation include being used as a means of controlling or removing invasive species. The system also can be used to increase the resistance of endangered species to diseases and other threats such as drought. An example of such use is targeting mosquitoes that spread avian malaria to birds and humans.

In agriculture, gene drives like those based on CRISPR can be used to make crop pests more susceptible to pesticides and herbicides. Crops can also be improved using the gene drives. The improvements include disease resistance, expression of valuable nutrients, and quality.

Other Applications of CRISPR in Crops and Livestock

The technology has been used to treat natural, parasitic, and vector-borne diseases in both animals and plants. The CRISPR-Cas9 system has been successfully used to genetically modify sorghum, wheat, maize, sweet orange, tomato, potato, duckweed, liverwort *Marchantia polymorpha* L., barley and *Brassica oleracea*, soybean, melon, and poplar (Cao et al., 2016). It also has been used successfully to genetically modify various organisms that include insects, fish, rabbits, pigs, mice, monkeys, and human cells (Mudziwapasi et al., 2018; Cao et al., 2016). The CRISPR-Cas9 system can be used to confer multiple pathogen resistance simultaneously. An example is a production of tobacco and *Arabidopsis* resistant to beet severe curly top virus, bean yellow dwarf virus, and tomato yellow leaf curl virus (Cao et al., 2016). CRISPR-Cas9 system can be used in the future to engineer completely new metabolic pathways, resulting in increased nitrogen fixation by engineered plants (reducing dependence on fertilizers), improved nutritious values, enhanced photosynthetic capacity, and plants that can be efficiently utilized in biofuel production (Lau et al., 2014). Using CRISPR, there is the possibility of making minimal plant cells (devoid of nonessential components and capable of division), similar to bacterial ghosts, which can be exploited as factories for novel biological systems (Noman et al., 2016). CRISPR-Cas9 was used to strengthen *Nicotiana benthamiana* plants' immunity against viral infections (Ali et al., 2015). It was also used to knock out three mildew-resistance loci (MLO) homolog alleles conferring heritable broad-spectrum resistance to powdery mildew in wheat plants (Wang et al., 2014). CRISPR array genotyping has applications in the identification and typing of bacteria at the strain level, where it has the benefits of being affordable and having a high resolution. Bacterial and archaea strains can also be differentiated by the presence or absence of CRISPR arrays (Barrangou and Horvath, 2012; Shariat et al., 2013). The limited presence of the CRISPR system in bacterial genomes, which is currently at approximately 46%, limits its applicability in this regard. However, CRISPR-based typing has been applied in spoilage organisms such as *Lactobacillus buchneri*, starter cultures such as *Salmonella thermophilus*, probiotics such as *Lactobacillus casei*, and foodborne pathogens such as *Escherichia coli* and *Salmonella* (Selle and Barangou, 2015). CRISPR-Cas systems can be heterologously introduced into organisms that lack endogenous CRISPR-Cas systems. This vaccinates the recipients against the uptake of undesirable genetic content (Sapranauskas et al., 2011). However, to use the CRISPR systems efficiently for gene therapy, there is a need to ensure immune stealth to minimize dosage and allow readministration of protein or nucleic acid therapeutics. This is because the cells' immune system can target CRISPR complexes for destruction if they are perceived as foreign.

CRISPR can be used to alter or control organisms that directly cause infections and diseases such as schistosomiasis or control organisms that are disease reservoirs such as bats and rodents. Gene drives such as CRISPR can be used to control or alter organisms that carry infectious diseases that threaten the survival of other species or eliminate invasive species that threaten natural ecosystems and biodiversity.

Despite CRISPR systems benefits and seemingly limitless applications, ranging from curing of genetic diseases to bring back species of organisms that are extinct, there has not been any use of the system in some African countries. Some African countries have not approved the use, cultivation, and or production of genetically modified organisms (GMOs). CRISPR genetically manipulated organisms may be approved easily since their products may not have transgenic footprints. This is because it delivers preassembled Cas9-sgRNA ribonucleoproteins or by transient expression of in vitro transcripts of the Cas9-coding sequence and sgRNA. Thus, it may not be regulated by the current biosafety regulations in some countries. Consequently, the common white button mushroom (*Agaricus bisporus*), modified to resist browning using the CRISPR-Cas9 system, was the first CRISPR-edited organism that received approval to be cultivated and sold without any further oversight of US regulations (Cao et al., 2016).

Other Applications of CRISPR in Humans

CRISPR-Cas9 can be used to investigate the treatment of cystic fibrosis (CF). Adult intestinal stem cells are obtained from two patients with CF and can be successfully corrected of the most common mutation causing CF in intestinal organoids. Once the mutation has been corrected, the function of the CF transmembrane conductor receptor (CFTR) is restored (Cao et al., 2016; Wojtal et al., 2016).

Another disease in which CRISPR-Cas9 has been investigated is Duchenne muscular dystrophy (DMD). AAV delivery of CRISPR-Cas9 endonucleases can be used to recover dystrophin expression in a mouse model of DMD by deletion of the exon containing the original mutation (Sapranauskas et al., 2011; Nelson et al., 2016). This produces a truncated but still functional protein. Treated mice will partially recover muscle functional deficiencies. Significantly, the dystrophin gene can be edited in muscle stem cells that replenish mature muscle tissue. CRISPR-Cas9 can also be used to treat hemoglobinopathies (Mudziwapasi et al., 2018; Cao et al., 2016; Selle and Barangou, 2015). BCL11A enhancer disruption by CRISPR-Cas9 can induce fetal hemoglobin in both mice and primary human erythroblast cells (Nelson et al., 2016; Noman et al., 2016). In the future, such an approach could allow fetal hemoglobin to be expressed in patients with abnormal adult hemoglobin (Mudziwapasi et al., 2018; Cao et al., 2016). This will represent a novel therapeutic strategy in patients with diseases such as sickle cell disease or thalassaemias (Noman et al., 2016; Selle and Barangou, 2015; Tabebordbar et al., 2016).

Advantages

CRISPR-Cas is relatively simple, inexpensive, easily programmed, and efficient. It has a high degree of fidelity (Kanchiswamy, 2016; Cong et al., 2013; Mali et al., 2013). The DSB generated by CRISPR-Cas9 may also lead to on-target mutagenesis effects, making genome modification highly efficient (Mahmoudian-sani et al., 2017; Steffan et al., 2018). Besides the types of insertions, deletions, translocations, and rearrangements, on-target effects include large chromosome deletions, chromosome truncations, and homozygosis of the genome by inter-homology repair (Shin et al., 2017; Cong et al., 2013; Pan et al., 2016).

CRISPR has relatively higher gene modification efficiency, and it is less time consuming because protein engineering is not required to have an effective endonuclease. Also, the gRNAs used in the CRISPR system are easy to design and customize to any target, as it only requires a 20 nucleotide guide sequence. Additionally, when using the CRISPR systems, multiple gene editing in the same cell can be achieved by introducing multiple gRNAs simultaneously (Cong et al., 2013; Mali et al., 2013). CRISPR can remove a gene without interfering with intracellular mechanisms (Mahmoudian-sani et al., 2017; Cong et al., 2013). Unlike ZFN and TALENS, which require dimers, CRISPR uses a single protein that cleaves both sites within the monomer.

A type II CRISPR system used from *Streptococcus pyogenes* is active in a range of species, and target sites are even easier to engineer, suggesting the system can be easily adapted to multiple vector species (Champer, 2016; Pan et al., 2016). The CRISPR system can be tweaked with activator and repressor domains to control gene expression rather than editing. Furthermore, systems based on Cas9 have shown the propensity to target heterochromatin sequences, targeting DNase-inaccessible locations. Cas9 systems tend to cleave more at highly methylated regions and are more flexible due to the multiple variants of Cas9 (Tang et al., 2019, Mahmoudian-sani et al., 2017; Champer, 2016). Prior screening for potential off-target sites to avoid off-target mutations reduces their occurrence.

In summary, CRISPR technology has led to increased understanding of genes, genome structure, and function. It has also given us access to models that were previously not accessible. CRISPR technology has greatly contributed in drug development and therapy (Robb, 2019).

Disadvantages

A major disadvantage when using CRISPR is how to deliver gene editing to the right cells, especially if the treatment is to be delivered in vivo (Kosicki et al., 2018). To safely deliver Cas9-nuclease encoding genes and guide RNAs in vivo without any associated toxicity, a suitable vector is needed, with AAV previously being a favored option for gene delivery. However, this delivery system may be too small to allow efficient transduction of the Cas9 gene (Peng et al., 2018; Schaefer et al., 2017).

The CRISPR system can be quite large, consisting of promoters, the Cas9 gene, and gRNAs. A system this size is prone to mutation and errors introduced during homing, including the potential loss of function of disease-refractory genes. CRISPR-aided genome editing can cause unintended mutations, which are commonly known as off-target mutations, into the genome. These can include single-nucleotide mutations as well as mutations in the noncoding regions of the genome and unspecific binding of gRNA causing unintended insertions at different loci. The mutations can be about 1,500 single-nucleotide mutations and even more than 100 larger deletions and insertions per single event. (Mudziwapasi et al., 2018; Nelson et al., 2016; Wang and Yang, 2019). Moreover, in-frame indel mutations could potentially generate gain-of-function mutations, the risks of which are even more difficult to predict (Stella and Mantoya, 2015; Kosicki et al., 2018). The unintentional edits of the genome could have profound long-term complications for patients, including malignancy (Schaefer et al., 2017; Stella and Mantoya, 2015).

However, these mutations may not be predicted using computer algorithms widely used to look for off-target effects but by whole genome sequencing (Peng et al., 2018; Schaefer et al., 2017; Stella and Mantoya, 2015; Chen et al., 2016). If the gRNA target sequence match with other sites in the genome, "off-targets" can occur (Xie et al., 2014). Resistance against CRISPR-Cas9 gene drives can develop because of indels introduced by NHEJ after an initial Cas9 cleavage event that can mutate the wild-type gene sequence and prevent the propagation of the drive (Mudziwapasi et al., 2018). It has already been shown that CRISPR-Cas9 technology can alter the genome of human embryos, which theoretically could prove useful in the preimplantation treatment of genetic diseases (Wang and Yang, 2019; Nelson et al., 2016; Mudziwapasi et al., 2018).). However, it has been noted that any genetic modification of the germline would be permanent, and the long-term consequences are unclear (Zafer et al., 2016). Editing early embryos using CRISPR does not only provide benefits for babies to render them immune from HIV, but poses potentially serious risks on multiple fronts (Kosicki et al., 2018; Cao et al., 2016; Wojtal et al., 2016). While many strategies for increasing HDR have been reported in cell lines, it is still questionable whether they apply to human embryos (Nelson et al., 2016; Wang and Yang, 2019).

Previous work done by Mitalipov's group suggests that the maternal allele could serve as a template for gene repair to achieve correction of pathogenic mutation, whereas other groups have argued that Cas9 may induce large-scale deletions or rearrangements that may lead to false-positive results using PCR-based genotyping (Kosicki et al., 2018; Zafer et al., 2016).

According to Ping (2017), the failure of CRISPR technology is also due to maternal effects. CRISPR gene drives also exhibit limitations of target loci due to restrictions of the PAM sites leading to variations in efficiency. In medicine, off-target activities can lead to the cleavage of the DNA; sequences that do not match the corresponding gRNA can therefore lead to pathologic conditions.

Bioinformatics tools are now used to predict the efficiency of the designed gRNA as indels like insertion, deletions, and addition or mutations can be generated as a DNA repair mechanism initiated by sgRNA guided cleavage of Cas9 (Swiech et al., 2015). The bioinformatics tools that can be used include SURVEYOR assay. Deep sequencing can also be used to verify the efficiency of the CRISPR system.

How to Counter Off-Target Effects

Mutations are likely to appear in sites that have differences of only a few nucleotides compared to the original sequence as long as they are adjacent to the PAM sequence. It occurs since Cas9 can tolerate up to 5 base mismatches within the protospacer region or a single base difference in the PAM sequence. CRISPR can be improved to reduce off-target mutations. This can be done through truncated gRNA within the crRNA derived sequence or by adding 2 extra (G) nucleotides at the 5 prime ends. This can also be achieved through paired nickases where D10A, Cas9, and 2 sgRNAs complementary to the adjacent area on opposite strands of the target site are used. This induces DSBs in target DNA and is expected to create only single nicks in affected target locations and therefore results in minimally affected target mutations. Cpf1 nucleases generate fewer off-target mutations or nickase/dimerizing nucleases that require the cooperative binding at two independent sites and may be able to initiate the homing reaction as compared to Cas9 (Tong et al., 2019; Hoffman et al., 2017; Gokcsezade et al., 2014).

Cas9n can be used in place of cas9 nuclease as they produce few off-target changes in the DNA (Cho et al., 2014). Three deep sequencing methods can be used to detect off-target cleavage that can occur during the use of CRISPR technology. The sequencing methods include high throughput genome-wide translocation sequencing (HTGTS), genome-wide unbiased identification of DSBs enabled by sequencing (GUIDE seq), and in vitro Cas9 digested whole genome sequencing (Mei et al., 2016).

6

Comparison of ZFNs, TALENs, CRISPR, and MegaTALs

This chapter compares the genome editing tools covered in this book, beginning with Figure 6.1 and Table 6.1.

Figure 6.1 is a diagrammatic representation of the four genome editing tools and how they assemble to effect cutting of DNA.

Table 6.1 shows a detailed comparison of the four genome editing tools.

Mechanism of Action

During genome editing, both ZFNs and TALENs utilize a nuclease (FokI) that requires dimerization to cut. Since the FokI cleavage domain used in ZFNs and TALENs is activated via dimerization, ZFN and TALENs require the construction and delivery of two protein halves. Each of the protein halves comprises a distinct DNA-binding domain that is fused to FokI, for a single target site. This architectural requirement limits the potential for ZFN and TALENs delivery into primary cells using viral vectors. It also limits their simultaneous (i.e., multiplexed) delivery of two or more nucleases to modify more than one gene at a time (Edgell and Stoddard, 2011). Unlike TALENs or ZFNs and megaTALs, CRISPR-Cas systems can be multiplexed. This is achieved by adding multiple different gRNAs to the target cells (Zuyong et al., 2016). Like TAL proteins, the CRISPR-Cas system can be tweaked with activator and repressor domains to control gene expression rather than editing DNA. ZFNs, TALENs, and megaTALs are more expensive than CRISPR, and they have more complex protein engineering than CRISPR. Each zinc finger module binds with a nucleotide triplet. TALEN subunits interact with single base pairs.

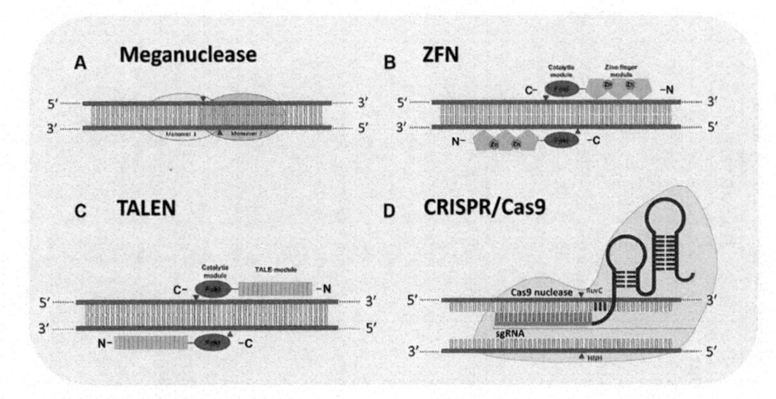

(Romay and Bragard, 2017).

Figure 6.1 Examples of the different gene-editing tools.

DOI: 10.1201/9781003165316-6

Table 6.1 Comparing the Gene-Editing Tools

	ZFNs XE "ZFNs"	**TALENs XE "TALENs"**	**CRISPR XE "CRISPR"**	**MegaTALs**
DNA XE "DNA" -recognition mechanism	Protein-DNA XE "DNA" interactions that introduce DSB XE "DSB"	Protein-DNA XE "DNA" interactions that introduce DSB XE "DSB"	RNA XE "RNA" -guided protein-DNA XE "DNA" interactions that introduce DSB XE "DSB"	Protein-DNA XE "DNA" interactions through HR
Recognition site	Typically 9–18 bp per ZFN monomer, 18–36 bp per ZFN pair	Typically 14–20 bp per TALEN monomer, 28–40 bp per TALEN pair	22 bp (20 bp guide sequence + 2 bp protospacer XE "protospacer" adjacent motif (PAM XE "PAM") for *Streptococcus pyogenes* Cas9 XE "Cas9"); up to 44 bp for double nicking	Between 14 and 40 bp
Use as a gene activator	Yes; activation of endogenous XE "endogenous" genes; minimal off-target XE "off-target" effects; may require engineering to target particular sequences	Yes; activation of endogenous XE "endogenous" genes; minimal off-target XE "off-target" effects; no sequence limitations	Yes; activation of endogenous XE "endogenous" genes; minimal off-target XE "off-target" effects; requires "NGG" PAM XE "PAM" next to the target sequence	No
Use as a gene inhibitor	Yes; works by blocking transcription elongation via chromatin repression; minimal off-target XE "off-target" effects; may require engineering to target particular sequences	Yes; works by blocking transcription elongation via chromatin repression; minimal off-target XE "off-target" effects; no sequence limitations	Yes; works by blocking transcription elongation via chromatin repression; minimal off-target XE "off-target" effects; requires "NGG" PAM XE "PAM" next to the target sequence	No
Target site length	18–36 bp per ZFN pair	28–40 bp per TALEN pair	19–22 bp	14–40 bp
Target specificity	High; G-rich sequence preference; Only some positional mismatches are tolerated; retargeting requires protein engineering	High; requires a T at each 5'-end of its target; some positional mismatches are tolerated; retargeting requires complex molecular cloning	Moderate; RNA XE "RNA" -targeted sequence must precede the 2 base pairs recognized by PAM; XE "PAM" only some positional mismatches are tolerated; retargeting requires new RNA guide; protein engineering is not required	High; only some positional mismatches are tolerated; retargeting requires protein engineering

Targeting constraints	Difficult to target non-G-rich sequences	5′ targeted base must be a T for each TALEN monomer	The targeted sequence must precede a PAM XE "PAM"	Targeting novel sequences often results in low efficiency XE "efficiency"
Off-target effects	Some mismatches are tolerated	Some mismatches are tolerated	Tolerant even of several consecutive mismatches	Some mismatches are tolerated
Ease of engineering	Difficult; may require substantial protein engineering	Moderate; requires complex molecular cloning methods	Easily retargeted using standard cloning procedures and oligo synthesis	Difficult; may require substantial protein engineering
Immunogenicity	Likely low, as zinc fingers XE "zinc fingers" are based on human protein scaffold; FokI is derived from bacteria and may be immunogenic	Unknown; protein derived from *Xanthomonas* sp.	Unknown; protein derived from various bacterial species	Unknown; meganucleases may be derived from many organisms, including eukaryotes
Ease of ex vivo delivery	Relatively easy through methods such as electroporation XE "electroporation" and viral transduction	Relatively easy through methods such as electroporation XE "electroporation" and viral transduction	Relatively easy through methods such as electroporation XE "electroporation" and viral transduction	Relatively easy through methods such as electroporation XE "electroporation" and viral transduction
Ease of in vivo delivery	Relatively easy as small size of ZFN expression cassettes allows use in a variety of viral vectors	Difficult due to the large size of each TALEN and repetitive nature of DNA XE "DNA" encoding TALENs XE "TALENs", leading to unwanted recombination events when packaged into lentiviral vectors	Moderate: the commonly used Cas9 XE "Cas9" from *S. pyogenes* is large and may impose packaging problems for viral vectors such as AAV, but smaller orthologs	Relatively easy as the small size of meganucleases allow use in a variety of viral vectors
Cost	High	High	Reasonable	High
Popularity	Low	Moderate	High	Low
Ease of multiplexing	Low	Low	High	Low

Source: Komor et al., 2017.

MegaTALs combine the easy-to-design DNA-binding domains from TALENs with the high cleavage efficiency of meganucleases in a single chimeric protein. The meganucleases cause less toxicity in cells when compared to ZFNs and TALENs. ZFNs, TALENs, megaTALs, and CRISPR have programmable site-specific nucleases. These can induce DNA DSB, stimulating the NHEJ or the HDR repair pathways at the targeted genomic loci.

TALENs can be designed to bind virtually any target sequence as long there is a thymidine at the 5′ end of the targeted sequence. MegaTALs are unique in that their DNA- binding and cleavage activities are integrated. The DNA target in megaTALs can be viewed as two distinct half-sites that are directly contacted through antiparallel β-sheets belonging to each domain of the pseudo dimer.

ZFNs, TALENs, and CRISPR are modular proteins. They interact with the major groove of the double helix structure of DNA to recognize specific base pairs. Unlike ZFN and TALEN platforms, the DNA catalysis active site in megaTALs is directly integrated into the DNA-binding interface. Thus, they do not need to be appended to a separate DNA cleavage domain in order to be used for targeted DNA-break based genome editing.

Nucleases

ZFNs and TALENs use FokI as the nuclease, while CRISPR uses Cas proteins. MegaTALs use a combination of TAL effector nucleases and meganucleases. Meganucleases (homing endonucleases) are single peptide chains. They have both DNA recognition and nuclease functions in the same domain. Transcription activator-like (TAL) effectors are DNA recognizing proteins. They are linked to separate DNA endonuclease domains for them to achieve a targeted DNA DSB.

DNA Target Recognition

The DNA target recognition of the nuclease platforms is governed by protein-DNA binding, except for CRISPR-Cas target recognition, which is primarily mediated via Watson-Crick base pairing. Target DNA recognition is governed by base-specific contacts made at the interface of the enzyme and DNA. MegaTALs, ZFNs, and TALENs contacts are made by amino acid side chains. CRISPR DNA recognition is accomplished through RNA to DNA base pairing in the guide sequence and base-specific amino acid contacts in the PAM region.

Recognition Site

The recognition site for ZFNs is 9–10 bp per ZFN monomer and 18–36 bp per ZFN pair. In TALENS, it is 14–20 bp per TALEN monomer and 28–40 bp per TALEN pair. For CRISPR, the recognition site is 22 bp. That is a 20 bp guide sequence plus 2 bp PAM for Cas9 and up to 44 bp for double nicking. The recognition site for meganucleases is 14–40 base pairs.

Specificity and Simplicity

MegaTALs are highly specific although they are complex to reprogram for the recognition of novel DNA targets. However, they are characterized by a lower on-target cleavage efficiency when compared to other genome targeting endonucleases. ZFNs, TALENs, and megaTALs tolerate small numbers of positional mismatches. CRISPR tolerates positional and multiple mismatches. CRISPR is more susceptible to off-target mutations. TALENs are highly specific because their cleaving activity requires dimerization of a TALEN pair before the FokI nuclease creates a DSB. However, CRISPR requires only one gRNA to make a DSB. CRISPR-Cas is the simplest of the genome editing tools. CRISPR, therefore, has a specificity disadvantage relative to TALENs.

TALENs DNA-binding domains have repeating units of 33–35 amino acids. Each of these contains two amino acids at defined positions. This confers specificity for a single nucleotide. There is a one-to-one relationship between these guiding amino acid pairs and the four DNA bases. This makes the design of new TALENs simpler than ZFNs. MegaTALs have higher rates of specificity than other available nuclease platforms. This is due to their large recognition sites, which are rare in any genome. This makes them useful in gene therapy, where only minimal off-target cuts are acceptable.

Targeting Constraints

In ZFNs, targeting a G-rich sequence is very difficult. TALENs require a 5' targeted base to be preceded by a T for each TALEN monomer. They also require a 13–18 bp space between FOK1 reverse pairs. In CRISPR, target sequences must be preceded by a PAM sequence. Engineering meganucleases used in megaTALs to target novel sequences usually results in low efficiency.

Ease of Engineering

Engineering in ZFNs and megaTALs is difficult, with TALENs engineering being moderately complex (Carroll, 2017; Beumer et al., 2013). CRISPR is easy to engineer. The gRNA design is based on complementarity with the target DNA. ZFNs assembled in arrays affect the specificity of neighboring ZFNs. This makes their design challenging (Marshall and Akbari, 2016). Their cloning varies; ZFNs require engineering linkages between zinc finger motifs, whereas TALEs do not require linkages. Unlike TALENs, CRISPR, and ZFNs, megaTALs cannot be used for gene inhibition. TALENs, CRISPR, and ZFNs block transcription elongation by chromatin repression. TALENs are not used frequently because they have single-site targeting, nonspecific mutations, and low efficiency. HDR can also be implemented to repair DSBs. However, it requires a homologous DNA template. This mechanism (HDR) cannot be used to implement a gene knocking into nondividing cells.

Immunogenicity

ZFNs have low immunogenicity because they are based on a human scaffold. However, FokI is derived from bacteria; hence, it may be immunogenic. Immunogenicity in CRISPR, TALENs, and megaTALs is not well known.

Ease of Ex Vivo Delivery

Ex vivo delivery of all genome editing tools is easy. It uses simple methods such as virus-mediated delivery and electroporation. gRNA used in CRISPR can be delivered to cells as a DNA expression vector, directly as an RNA molecule, and as a pre-loaded Cas9-RNA complex.

Ease of In Vivo Delivery

In vivo delivery is easy in ZFNs and megaTALs due to the small size of their expression cassettes. This allows them to be used in a variety of vectors. MegaTALs have an 8–12 bp repeat TAL array that is appended to the N terminus. This makes them ~2.5 kb per monomer. TALENs are the largest of genome editing tools, at ~3 kb/monomer (~6 kb/pair). Their highly repetitive sequences make their delivery using commonly used viral vectors difficult. They thus lead to unwanted recombination events when they are packaged into lentiviral vectors. CRISPR complex delivery is moderately easy. The large Cas imposes packaging problems in viral vectors. The originally adapted CRISPR-Cas9 from *S. pyogenes* is too large to fit into an AAV with a promoter and a gRNA. However, new Cas9 orthologs smaller in size have been discovered. These include the 3.2 kb Cas9 from *Staphylococcus aureus* (Sebastiano et al., 2011).

Cost Effectiveness

CRISPR is the most cost-effective genome editing tool. It is 3–6 fold cheaper than TALENs per reaction. TALENs are more cost effective for gene silencing as compared to knocking in/knocking out a gene.

Speed

CRISPR is the fastest and most straightforward way for genome editing.

Efficiency of Gene Editing

MegaTALs, ZFNs, TALENs, and CRISPR achieve efficient gene editing above 70% when in cell lines with high-level delivery and no toxicity.

Biology at the Breakpoint

MegaTALs leave 3' overhangs while ZFNs and TALENs leave 5' overhangs. The CRISPR-Cas9 system leaves blunt ends. This can result in the recruitment of different DNA repair proteins. Furthermore, this can change the expected outcome.

Table 6.2 shows a relative comparison between each platform for various salient designer endonuclease parameters.

Table 6.2 A Relative Comparison of ZFNs, TALENs, CRISPR, and MegaTALs

	megaTAL	ZFN	TALEN	CRISPR Cas9
A Engineerable unit	Integrated	*Pseudo* modular	Modular	Modular
B Coding sequence size	1,000 bp/ 2,500 bp	1,000 bp/ monomer	3,000 bp/ monomer	4,200 bp + gRNA
B Architecture	Monomer	Dimer	Dimer	Protein RNA
C DNA binding	Protein	Protein	Protein	RNA/Protein
C Catalytic specificity	High	Low	Low	Medium
D Overhang	3′	5′	5′	Blunt
D Multiplex compatibility	Medium	Low	Low	High

NB. A = Development, B = Delivery, C = Specificity and D = Efficiency

Source: Mei et al., 2016.

7

Gene Drives

Introduction

In 1950, after Hermann Muller received the Nobel Prize for radiation-induced genetic mutations, Edward Knipling, an entomologist, wrote to Herman Muller. He was asking whether pest insects that become genetically sterile when mutated by exposure to radiation could be used to eradicate destructive species. This steered the development of the sterile insect procedure, which was used to eradicate a major cattle pest from nearly all of North and Central America. This saved billions of US dollars and lessened the use of insecticides.

Using this technique, hundreds of millions of irradiated insects were reared and released (Brossard et al., 2019). This method only worked on a few pest species. This led researchers to reflect on how it could be developed to widen its reach. The strategy appeared to be developing a means by which genetically manipulated insects could be released with genetic traits that spread over a population, although they conferred no value to the insect.

Fast-forward to the era of genomics, and we seem to be on the verge of genetic methods that will allow us to spread genes that disrupt the capacity of pests to transmit pathogens or bias their sex ratio so that essentially all offspring are male. Naturally, a population with few females is destined for extinction. There are naturally two constituents to these manipulations (Brossard et al., 2019). First, there must be a way to modify or deactivate a gene to change individual pest's biology. Then there needs to be a technique to link it to a gene or to genetic systems to drive it into the population (Fears and ter Meulen, 2018).

Some genes persist and spread down generations and throughout a population (Eidgenossenschaft et al., 2018). This is because they have a useful function for the organism. This includes reproduction and survival. They thus spread in the population by positive Darwinian selection. The genes are maintained by purifying selection in the face of recurrent mutation. However, some genes persist by distorting their generation to generation transmission. They can thus be transmitted to more than 50% of the gametes of an individual. This can occur even though the individual can have inherited the respective gene from just one of the two parents (AAS, 2017).

Gene transmission thereby increases the respective gene's frequency between generations, giving the gene an advantage. Zygotic and gametic killers, meiotic drivers, homing endonuclease genes (HEGs), transposable elements, and B chromosomes are some of the genes or genetic elements that show such distortion of the transmission ratio or "drive" (Burt and Crisanti, 2018). Gene drives are thus biological mechanisms that accelerate the transmission of genes (Thompson, 2018; Macias et al., 2017). They result in the passing on of gene variants with much higher inheritance rates. These rates are higher than provided for under Mendel's laws of inheritance (Rode et al., 2019; Vlugt et al., 2018). Gene drives result in the preferential increase of a specific trait in subsequent generations and possibly in the population (Rode et al., 2019). These genetic elements have evolved many times in different taxa. Drive affects genomic features that include base composition, genome size, the shape of a chromosome, repeat structure, centromere structure, and the distribution of recombination hotspots.

Different molecular mechanisms underlie gene drives. However, in some of the mechanisms, gene drives can copy themselves onto opposite chromosomes (homing). This causes most or all offspring to inherit the respective gene drive allele. Gene drives can also reduce the viability of the gametes inheriting wild-type alleles. This gives a fitness disadvantage to the wild-type alleles as compared to the gene drive allele. Consequently, most or all the offspring inherit the gene drive allele (Champer et al., 2016).

DOI: 10.1201/9781003165316-7

Some driving genes can spread although they do not have a useful function for organisms that carry them. Additionally, some driving genes can harm the organisms carrying them. These scenarios occur when the transmission distortion is greater than the effect of reduced survival and reproduction. These genetic elements are thus called selfish genetic elements or selfish genes. Since the selfish genetic elements can be harmful to the organisms carrying them, the genes that can suppress them spread via natural selection. This occurs in a way analogous to the spread of genes that suppress parasites. The selfish genetic elements are selected to avoid these suppressive genes, potentially leading to an arms race. At times, some features of selfish genetic elements are co-opted to perform a function useful for the host organism. This includes antibody diversification in vertebrates, mating type switching in yeast, and telomere maintenance in *Drosophila* (Burt and Crisanti, 2018).

History of and Principle Behind Gene Drives

The utilization of natural gene drives in spreading genetic variants in a population was first conceptualized over 50 years ago. To date, some technical difficulties are still being encountered in creating working systems based on these mechanisms (Norwegian Biotechnology Advisory Board, 2017). Consequently, natural selfish genetic elements inspired the development of synthetic gene drives. The natural genetic elements are found in some single-cell organisms, and an example is homing endonuclease genes (HEGs) (Hammond et al., 2016).

Synthetic gene drives have wide applicability, although their development has been very modest in the past decades (Champer et al., 2016). The development of CRISPR, a genome editing tool, around 2012 sparked more interest in gene drives (Spaak, 2019; Norwegian Biotechnology Advisory Board, 2017). This is because CRISPR can be used to engineer gene drives with relative ease. Synthetic gene drives had remained largely theoretical before then (Evans and Palmer, 2018). Proof of the potential of CRISPR-Cas gene drives came around 2015. This was shown in research on fruit flies. Use of a gene drive to completely wipe out caged mosquitoes after between 7 and 11 generations provided more evidence of their capabilities. The gene drive destroyed a sex-determination gene that was essential for the survival of the mosquito (Thompson, 2018). Thus, genome editing tools such as ZFNs, TALENs, megaTALs and CRISPR helped turn gene drives from being theoretical to a practical principle (Esvelt, 2019). Gene drives can now be used to introduce genetic information for new traits into a population.

Gene drives have two key classes of application, which have implications of different significance. These include:

1. Introducing a genetic modification in laboratory populations. Once a strain or a line carrying the gene drive has been produced, the drive can be passed to any other line by mating. Here the gene drive is used to achieve much more easily a task that could be accomplished with other techniques.
2. Introducing a genetic modification in wild populations. Gene drives constitute a major development that makes possible previously unattainable changes.

Classifications of Gene Drives

Gene drives can also be referred to as active genetic elements (Grunwald et al., 2019). Drive mechanisms can be categorized into two main types, which attempt to achieve inheritance bias through:

1. **Over-replication** of the genetic element, also referred to as a replication distorter.
2. **Preferential segregation** or transmission of the genetic element, also referred to as a "transmission distorter".

CRISPR makes use of both the over-replication and preferential segregation mechanisms. This makes it possible to come up with tailor-made drives for the distinct mode of action. However, overlapping can also occur among the different mechanisms, as diverse as they are. There are several gene drive classes, described in the following sections.

Standard Drive Systems

Standard drives can spread to populations connected by gene flow. They affect large populations and cost less to deploy. They include transposons that duplicate themselves.

Local Drive Systems

Local drive systems can spread through regional populations. However, they cannot spread to all populations that are connected by gene flow. Subtypes of local drives are the self-exhausting drives. Local drive systems are suitable where geographic and temporally limited interventions and expedient field tests are required. These can be for the determination of the ecological effects of more powerful drive systems. They include killer-rescue systems. These use independent toxin and antitoxin genes in spreading cargo genes that are associated with the antitoxin.

Daisy Drive Systems

Daisy drives separate components of an endonuclease drive system into daisy elements. These are arranged in a linear daisy chain or parallel daisy field structure. They rely on progressive loss of the non-driving elements to limit the total spread of the cargo element.

Threshold Drive Systems

Threshold drives cannot be used where there is a large wild population. However, they are well suited to small islands or subpopulations that have a minimal gene set. These can include chromosomal translocations and *Medea* drives that are native to the *Tribolium* beetle.

Combination Drive Systems

Combination drives include the daisy quorum system. The daisy quorum system combines a self-exhausting daisy drive with a threshold dependence. This makes it able to efficiently alter a whole local population without invading other nearby populations. Combination drives can vary in their degree of reversibility. They also vary in their ability to maintain costly alterations and their capacity to suppress and alter populations.

Natural and Synthetic Gene Drives

There are natural and synthetic gene drives. Natural gene drives occur naturally, while synthetic gene drives are created in laboratories, typically by genetic modification (Vlugt et al., 2018). The natural gene drives are typically selfish genetic elements. They have a biased inheritance leading to their super-Mendelian inheritance. This gives them a higher chance of being passed on to the offspring (Simon et al., 2018). This chance can be as high as 100%. Natural gene drives include transposable elements, meiotic drives, and homing endonuclease genes (HEGs). Synthetic gene drives typically influence target populations via population alteration or population suppression. Population suppression gene drives reduce the total number of individuals in a population over generations following their introduction. The number of individuals can end up at zero. A population alteration gene drive alters some characteristics of a population following its introduction. The alteration can confer immunity or resistance to diseases or parasites. The current genome editing tools such as CRISPR can be used to create synthetic gene drives in a short time and in a targeted way (Liddicoat, 2016).

Suppression Drives

Drives can be suppressing or modifying. Suppression drives aim to reduce or eliminate a population. In combination with an independent threshold surge, they can spread globally and lead to the eradication of a whole species. This can be the case in particular if the training system itself has low intrinsic fitness costs and no resistance develops during the training (Champer et al., 2016).

Modification Drives

In a modification drive, the changed allele is transmitted by "hitchhiking" with the drive to wild-type chromosomes (Hammond et al., 2017). The replication of the drive allele may happen in offspring in the initial zygote stage. If this occurs, all of the offspring's cells will be homozygous for the changed gene (Enzmann, 2018). Then again, the drive might only be active in the offspring's germline. In this situation, all gametes will have the drive allele, but cells that make up other tissues will stay heterozygous.

Modifications are intended to spread certain characteristics in a population, for example, with using payload genes to merely alter and cause a change that aims to be proliferated to the subsequent generations. If this type of drive overcomes the problems of resistance, it can in turn genetically change a whole species. Modification drives may be described as replacement drives, although this term can give an off-context impression in implying that an entire species can be removed and another species takes over in its place.

Threshold Dependent Drive

Fixation is the major goal of this drive, and it only proliferates within a population. However, it is worth noting that the proliferation is controlled by a threshold frequency below which the drive will just be eliminated (Min, 2018). It is argued whether or not the predictions of simple models are valid in the wild population and also whether or not the dispersal methods and organism in question have any implications (Marshall and Akbari, 2018).

Threshold-Independent Drive Systems

This drive is not limited by threshold, as the name implies. It can successfully proliferate and establish itself without a minimal frequency to adhere to. This makes it very much able to spread quickly in a population group as well as give it high invasiveness. Due to its nature, it can greatly impact the environment and biodiversity at large.

Temporarily Self-Limiting Drives

This is the category under which the famous daisy chain drive falls. It is deemed self-limiting in that from the way it is designed, the drive will automatically stop after a certain set number of generations. This can be achieved by incorporating elements that obey Mendel's segregation laws.

Selfish Genetic Elements

Genes within a genome are commonly seen as working together collaboratively to produce a viable organism (Runge and Lindholm, 2018). In this collaboration, all the genes have the same probability of being inherited and expressed in the progeny, so the offspring inherits equal halves of its genetic makeup from both parents, as most organisms contain two sets of genetic material commonly known as chromosomes, a scenario described as diploid. Every single gene for a specific trait is therefore present in two copies occupying the same locus on the homologous chromosome. The two copies of the same gene are called alleles. They can code for a particular trait such as height and will be passed on to future generations on a 50% probability. However, the selfish genetic elements in question here do not play by the 50% rules and so are described as selfish in that they only seek their interests.

They are also capable of altering the odds of inheritance in their favor and thus can rapidly propagate through populations (Manser et al., 2017). Selfish genetic elements do not regard at all what implications their actions may have on the host organism, even if it means compromising the well-being of the host organism. It is only natural for the organism to react defensively, and this creates selection pressures on the drive element.

Selfish genetic elements may be whole genes or just chromosomes; for example, B chromosomes have unique ways of ensuring their persistence in the progeny. They can even eventually be expressed in an entire population. According to McLaughlin and Malik (2017), over-replicators are selfish genetic elements that proliferate by duplicating themselves to other parts of the host genome, known as transposable elements. Another group of interest is termed transmission distorters, which disrupt the normal transmission mechanisms and ensure that they are the only genes passed on to the next generations, at the expense of the rest of the genes.

Over-Replicators

Over-replicators shift odds in their favor by creating extra copies of themselves in the genome. They may be transmission distorters. They achieve an inheritance bias in their favor by creating extra copies of themselves in the genome. There are two members in this group, the transposable elements and HEGs.

HEGs (Homing Endonuclease Genes)

These are genes coding for an endonuclease enzyme that can cleave DNA at specific sites, ideally below 40 base pairs, and successfully copy themselves in the middle of that sequence through homology repair. This process is known as homing. This means that the homing gene becomes present within the recognition sequence that the endonuclease has to cleave. HEGs are present in many bacteria, fungi and plant cells. When the HEG is present in a single chromosome and absent in the homologous counterpart, the endonuclease detects the recognition site on that chromosome and induces a site-specific DSB, ensuring that the HEG is copied via a homology-directed mechanism.

Segregation and Transmission Distorters

Sex-Ratio Distorters

Sex-ratio distorters are also known as sex-linked meiotic drives. They function mainly by biasing the gender ratio in such a way that the larger fraction of offspring is inevitably a particular gender, for example 90% males and 10% females (Champer et al., 2016). They seem to have taken the spotlight in the development of synthetic gene drives. They are highly applicable in population control, where the increase in population size may be determined by how many males or females are present. Skewing the sex ratio by increasing the number of males and reducing females can reduce population size. This is because males are incapable of carrying progeny on their own, and female offspring will be gradually declining with each generation. Eventually, that population will cease to exist. Such a drive may be introduced into the male chromosome or any other chromosome that is not sex-linked. The highest and swiftest suppression rate can be achieved if the drive is linked to the Y-chromosome (Marshall and Akbari, 2018). Sex-ratio distorters gain an advantage by eliminating the competition.

Toxin–Antidote Based Drives

In this mechanism, a toxin and antidotes work together to achieve an inheritance skewing. Simply speaking, the gene is introduced as a toxin and another gene is made available as a counter-gene and acts as the antidote (Brandt and Bucking, 2019). This antidote gene will, however, not be silenced by the mechanism in play. Alternatively, the construct can be designed such that a toxin protein's effects will be undone by an enzyme or RNA-based silencing mechanism. However, the toxin must create such deleterious effects that it can only be redeemed by the antidote, and if the antidote is absent, the organism will die. For this to be successful, the toxin and the antidote gene must be separated so that their inheritance together is not obvious, or they can be separated using a time frame space between expression and also developing a long viability product. A gene of interest can be bound to the antidote gene ensuring its expression and inheritance, or simply drive.

Maternal-Effect Dominant Embryonic Arrest (MEDEA)

The engineered *Medea* systems behave as modification drives by utilizing an RNA interference (RNAi)-based toxin–antidote combination. As indicated by Ward et al. (2011), the toxin consists of a microRNA (miRNA) that is expressed during oogenesis in *Medea*-bearing females, disrupting an embryonic essential gene in all embryos, regardless of whether those embryos have inherited a *Medea* or wild-type allele from the mother. A tightly linked antidote consisting of a recorded version of the target gene that is immune to the effect of the miRNA is expressed at the zygotic stage early in embryogenesis, only in those embryos that inherit the *Medea* element. This creative combination of maternally expressed toxin and a zygotically expressed antidote results in the survival of 50% of the embryos originating from a *Medea*-bearing heterozygote female, as those that fail to inherit the *Medea* element perish. Moreover, if the female has mated with a *Medea*-bearing heterozygous male, the antidote from the male will also take effect in the embryo, resulting in 75% of the embryos surviving. Consequently, *Medea* possesses a frequency-dependent fitness advantage compared to chromosomes lacking *Medea*, allowing it to rapidly drive a linked payload gene through a population, and it can be used for population suppression (Akbari et al., 2014; Marshall and Hay, 2011).

Synthetic *Medea* elements, consisting of a microRNA that targets and silences a maternal gene necessary for embryonic development (maternal toxin) linked to a zygotic antidote gene that rescues that function, have been inserted in the *Drosophila melanogaster* genome using the P-element transposon (Akbari et al., 2014; Chen et al., 2007). They are predicted to spread from low frequencies (Ward et al., 2010), an encouragement for the population replacement strategy. However, they raise the possibility that *Medea*-linked transgenes may spread into countries without their consent (Marshall, 2010). Consequently, there is interest in gene drive systems that, while strong enough to bring about population replacement at an isolated release site, are unable to establish themselves in neighboring populations. Hence, a novel gene drive system was proposed, inverse *Medea*, which displays these properties.

One major limiting factor that has impeded their development is the lack of a basic understanding of how to achieve effective RNAi-mediated silencing of key genes in the germline of species other than *D. melanogaster*.

Genetic Underdominance

Underdominance, also known as heterozygote inferiority, occurs when heterozygotes (or their progeny) have lower fitness than parental homozygotes. Underdominant systems function as a bistable switch: even when two underdominant traits confer equal fitness in the homozygous state, the one with the lower initial frequency will be lost in a large interbreeding population, whereas the other will spread to fixation (Davis et al., 2001). This occurs because individuals bearing the less common alleles are more likely to mate with the opposite type and produce unfit offspring, reducing the likelihood of the less common allele being passed on to future generations.

Consequently, underdominant systems require a high introduction threshold to spread through a population, they are likely to be confined to a local area, and they can be removed by inundation with wild-type organisms. Strategies to engineer underdominant gene drives using combinations of toxins and antidotes have been proposed and implemented in the fruit fly *D. melanogaster*, as fully functional systems capable of invading wild populations (Akbari et al., 2013). Developing small RNA-based toxins could prove difficult. An alternative approach may be to use RNA-guided endonucleases targeting mRNA as a toxin element in underdominance.

Transposable Elements

Transposable elements (TEs) are also referred to as transposons or jumping genes. They are small DNA segments that can move from one part of the genome to another by excising themselves and randomly inserting elsewhere in the genome. In the context of a gene drive, TEs typify an over-replication mechanism. Multiple copies of the same TE often amass in the genome (i.e., increase in copy number) due to DNA repair or gene replication mechanisms that operate in eukaryotic cells. Thus, the copy number of TEs typically exceeds what would be expected under Mendelian inheritance.

In 1952, Barbara McClintock discovered TEs, she observed that some DNA sequences in maize could occasionally change their location in the genome and suggested these "controlling elements" could potentially turn genes on and off (McClintock, 1956). Since then, scientists have found that TEs are ubiquitous among eukaryotes and often constitute a major part of the genome (Wicker et al., 2007).

The P-element transposon has long been used to create genetically modified *Drosophila melanogaster* in the laboratory (Rubin and Spradling, 1982). Meister and Grigliatti (1993) first showed that a P-element transposon could rapidly spread a specific gene into an experimental *Drosophila melanogaster* population. Although P-elements are specific to *Drosophila melanogaster*, the piggyBac and Hermes TEs have been used for transformation in mosquitoes with varying degrees of success. One of the most prominent is the use of transposons linked to a genetic payload, which would increase the frequency of the transposable element and genetic payload in the genome of a target organism, and eventually in the population (Rasgon and Gould, 2005). The use of TEs as vectors for a gene drive has several disadvantages. This includes insertion into random locations, relatively low transforming frequency, limited cargo gene size, and low stability of the integrated sequence (Fraser, 2012). Smith and Atkinson (2011) also indicated that transposable elements often have transposition rates that are too low to be usable, are unpredictable owing to lack of control over their integration sites, and have proven to be difficult to mobilize after integration.

Semele System

The *Semele* system, conversely, uses a paternal semen-based toxin and a maternally delivered antidote. *Semele* is a single-locus system consisting of a toxin gene expressed in the semen of transgenic males that either kills or renders infertile wild-type females and an antidote gene expressed in females that protects them against the effects of the toxin. If only males carrying the *Semele* allele are released into a wild population, they are expected to suppress the population size when released in large numbers. This happens because all of the wild females that mate with the *Semele* males are susceptible to their toxic semen. If both males and females carrying the *Semele* allele are released, the system displays bistable dynamics with a threshold frequency of about 36% in the absence of fitness costs (Marshall, 2011).

Merea System

The Merea system functions similarly to *Medea*, but the antidote to the maternal toxin is recessive.

Medusa System

Medusa is a two-construct design within the toxin–antidote system. It again is simply a theoretical design intended for population suppression, in which a population crash might be kept to geographical limits (Marshall and Hay, 2014). Medusa is made up of four components, two toxins, and the two respective antidotes. One toxin and the antidote will be located on the X-chromosome; the second toxin with the antidote to the first will be located on the Y-chromosome. One without the presence of the other could therefore not survive. This system will thus select for individuals with both the transgenic X and Y chromosome, thus selecting against females (XX), and if initially released at a sufficiently high frequency, could bring the population to collapse.

T-Complex or T-Haplotype

The t-haplotype or t-complex is a meiotic drive and sex-ratio distorter located on chromosome 17 that naturally occurs in mice. Its discovery goes back to 1927, when Nadine Dobrovolskaia-Zavadskaia, evaluating X-ray experiments in mice, first thought this to be the gene for short tails or taillessness. However, further evaluations with crosses showed that "tailless mice produced only tailless litters upon intercrossing, but neither short-tailed nor normal-tailed pups. Inspection of the embryos from such crosses showed that about half of the embryos died in utero (Herrmann and Bauer, 2012).

It was much later that it became evident that this region of chromosome 17 was what would later be called a selfish genetic element, containing not only genes for transmission distortion but also for male infertility and embryonic lethality. Mice that are homozygous for the t-complex will die before birth, and males with a copy of the t-complex will pass this on to 90% of offspring (Lindholm et al., 2013). This lethality is based on a toxin–antidote system, where a toxin will be released into the cells during spermatogenesis, and only those sperm will survive or be able to fertilize an egg cell that carry the gene for the antidote, which is located on the t-complex.

This meiotic drive system is specific to mice. The idea here is to create "daughterless" mice by modifying the t-complex with a mouse gene called Sry.24. This gene will act during embryo development and trigger the development of male characteristics irrespective of the actual gender of the mouse. Released into the wild, any offspring should have male characteristics, with no females left to breed, to collapse the population.

Translocations

Translocations result from the mutual exchange of chromosomal segments between nonhomologous chromosomes. Translocation heterozygotes are usually partially sterile, while translocation homozygotes are usually fully fertile. This effect is manifest during meiosis, when nearly half of the gametes from a translocation heterozygote have a duplication of one chromosomal segment and a deficiency of another. The haploid gametes are functional, but when they fuse with native gametes following fertilization, the resulting zygotes are inviable. This produces the bistable dynamics described for other under dominant systems.

Application of Gene Drives

Gene drives can be applied in fields such as agriculture, ecosystem conservation, public health, and basic research. They can be used to control or alter organisms carrying infectious diseases such as malaria, dengue, Chagas (American trypanosomiasis), and Lyme disease. Gene drives can also be used to alter or control disease vectors and other organisms that act as reservoirs of disease, such as rodents and bats (Heitman et al., 2016; Champer et al., 2016).

In ecosystem conservation, gene drives can be used to control or alter disease-carrying organisms threatening other species' survival. These include sleeping sickness and malaria vectors.

The vectors can be manipulated to make them unable to carry the disease parasite. If this feature can be spread across the entire population of the vector, then the vector-borne diseases can be eradicated (The Royal Society, 2018). Another application is to make threatened animal and plant species more resilient to disease and other threats in nature that are fighting against their existence or their proliferation for future generations (Rode et al., 2019). This is called genetic rescue.

Gene drives can be used to eliminate invasive species threatening ecosystems and biodiversity and endangered or threatened species. In agriculture, gene drives can be used to control or alter crop-damaging organisms and crops that carry diseases and to eliminate weeds. Gene drives can also be used in research to study species biology and mechanisms of disease.

Gene drives can spread beneficial traits through populations and provide all desired modifications within a population, though for an indefinitely long time. The ability to carry out cycles of modification that create and then leave behind a minimal genetic footprint while entering and exiting a population provides important points of control. This then helps in the replacement of broken elements, upgrades with new elements that better carry out their tasks, and/or provides new functions, all while promoting the removal of modifications no longer needed or that have served their purpose.

The gene drives offer a better way to eliminate rodents instead of current methods such as poisoning and capture. They can also be used to add desirable properties to organisms at the embryonic stage or even in the egg cells so that an organismal change can be possible. This can help a population's ability to improve by improving resilience to environmental changes and invasive parasites. The expected benefits compared to conventional approaches would increase the reproductive level of the positive genes and increase the ability of the populations to become future-proof through the introduction of adaptation traits before the arrival of the selective threat (The Royal Society, 2016).

Advantages of Gene Drives

1. Faster than natural mutations.
2. Guaranteed expression in progeny species.
3. Not limited by Mendelian laws.
4. Applications are not limited to one field but cut across all fields from agriculture to the health field and the environmental field.
5. Less labor intensive and more friendly, unlike current methods in use, for example, pesticides and herbicides or trapping and poisoning for rodents.
6. Less costly in general.
7. Using gene drives for population management could have lower health, economic, and environmental costs than traditional control methods. Gene drive-mediated pest control can therefore be very attractive for agribusiness because it allows direct manipulation of pest species. This is more complicated to achieve with classical GMO technologies.
8. Gene drives can easily eradicate a species, and large effects are expected within just a few years after the release of the first gene drive organisms into the wild (Courtier-Orgogozo et al., 2017).

Disadvantages of Gene Drives

1. Safety is not guaranteed. Gene drives can introduce a trait into a species and this can persist into the environment possibly causing deleterious effects to the environment. Since this is a new area, there are no reliable means to predict the probability and severity of the adverse events. So it is a risky undertaking with unsecured results.
2. Too many factors are at play with technology, so replicability of the results across different species may be a problem.
3. As amazing as this technology is, it can also be used for bioterrorism attacks, which are worse than cyberattacks because millions of lives can be lost in a very short time.
4. There are many ethical questions about the possibility of humans to change and to choose the characteristics of their children. This will mean the superiority of some people over

others. Moreover, the gap in the social strata may only become greater, as only the wealthy might have the means to use the technology.
5. There is a risk of failure of countermeasures to end a continuous spread of the gene drives (gene drives escaping control). They might enter nontarget populations, jump between species, and have unintended negative or positive consequences.
6. There can be resistance to the gene drive that can either occur at the molecular level when a chromosome is not recognized or cleaved by the Cas9 enzyme or at the behavioral level when wild-type individuals avoid mating with gene drive individuals.
7. There can be molecular off-target mutations (Nash et al., 2019; Hammond et al., 2018; Giese et al., 2019; Esvelt, 2017).

Concerns with Gene Drives

There are concerns that population suppression gene drives might result in the population or species becoming extinct (Esvelt, 2019). Gene drives can change a population's gene pool. Consequently, some genetic information will prevail within an entire population. Other nontarget populations can be unintentionally affected or eradicated in cases where the genetic information entails a lethal factor (Eidgenossenschaft et al., 2018).

Additionally, the safety of the environment can be compromised when synthetic gene drives are unintentionally released into the environment (Rode et al., 2019). The results of releasing synthetic gene drives or using them to eliminate a whole species from the environment are unknown. Some organisms can become resistant to the gene drive. The organisms carrying the synthetic gene drives can cross national borders into countries where they might be illegal. This makes their control difficult and their effects widespread (Heitman et al., 2016). As a result, conflicts can erupt between neighboring jurisdictions. There are also concerns that risk assessment frameworks currently in place are inadequate. This puts wildlife and other life forms at very high risk (Simon et al., 2018). This is especially so because some gene drives can spread to all target species populations connected by gene flow. This is compounded by the potential of some gene drives to become invasive (Esvelt, 2019). Another concern is the potential of gene drives to affect nontarget species. This might happen where there is a possibility of interspecies reproduction. This includes the hybridization of plant species that are closely related (Norwegian Biotechnology Advisory Board, 2017). The effects of unsuccessful suppression gene drives can reduce genetic diversity in the population of the target species, thereby affecting its fitness in difficult-to-predict ways (Simon et al., 2018). Resistance to the gene drives and off-target effects are also some concerns. Their effects are not very predictable and can reach diverse wild populations (AAS, 2017). Lastly, there are concerns about biosecurity issues concerning the use of gene drives. Some people think they can be used to deliberately transmit harmful pathogens to humans and animals. They can also be used to damage food sources. Thus, there is a need for strict control and monitoring of applications of gene drives.

Regulatory Systems

Gene drives are regulated as a way of controlling or governing them. This includes regulating their use, safety and spread (Liddicoat, 2016). Some countries do not have sufficient regulations for gene drives and end up just applying the ones for any other biotechnology. This is because synthetic gene drive organisms (GDOs) are products of genetic modification and are therefore subject to GMO national and international regulations. However, regulations for gene drives and GDOs should deal with the maximum implications and potential impacts of gene drives (Esvelt, 2019). The New Partnership for Africa's Development (NEPAD) agency was formed to promote the safe exploration and harnessing of advances in science-based innovations. This includes modern biotechnology. It helps member states to implement functional regulatory frameworks for the innovations. NEPAD then established the African Biosafety Network of Expertise (ABNE), the African Medicines Regulatory Harmonization (AMRH), and the African Biosafety Network of Expertise (ABNE) (Glover et al., 2018). ABNE helps African Union member states to build functional regulatory systems for safe and responsible application of agricultural biotechnology.

The Cartagena Protocol on Biosafety regulates GMOs at the international level. It is an agreement under the UN Convention on Biological Diversity (CBD). The protocol regulates GMOs' trade. It also seeks to protect biodiversity and health during the transport, handling, and use of GMOs (Norwegian Biotechnology Advisory Board, 2017). However, some countries have not ratified it (Rode et al., 2019). The UN Convention (1977) on the Prohibition of Military or Any Other Hostile Use of Environmental Modification Techniques (ENMOD) and the UN Convention (1972) on the Prohibition of Biological Weapons (BWC) can also regulate military use of gene drives *inter alia* (Norwegian Biotechnology Advisory Board, 2017).

8

Supernumerary B Chromosomes

Introduction

The word "supernumerary" is a term portraying repetition, the presence of something as an extra that is subsequently not required. In essence, we are stating B chromosomes exist notwithstanding the typical karyotype normally known as the A chromosomes (Ahmad and Martins, 2019). The B chromosomes in this way have no essential job in the life of a species. The living life form can thus manage without them, yet that does not mean they are useless. B chromosomes or embellishment chromosomes, as certain writings would depict them, have the distinct highlight of not following or not agreeing to the Mendelian isolation laws (Houben, 2017). Most researchers depict B chromosomes as egotistical and parasitic in that they utilize the cell's assets, for example, a cell apparatus that is required for the legacy and multiplication of the standard supplement set of chromosomes.

Lowell Randolph in 1928 is the person who named the supernumerary chromosomes B chromosomes, while he named the essential set A chromosomes, which is the way we recognize the two sets to date (Ahmad and Martins, 2019). Conversely, B chromosomes in different creatures display drive, so they carry on in explicit manners that challenge typical Mendelian transmission designs to guarantee their legacy in ensuing ages. As already explained, B chromosomes can defy Mendelian laws of segregation, and this gives an exceptional ability to inherit their heritage to future generations. B chromosomes are an important tool and have an important role to play in genetic engineering. This is because these supernumerary chromosomes can facilitate the migration of whole genomes and promote the sustainable development of genes of interest to cells. The B chromosomes are therefore key players in genetic engineering. There are many reasons why scientists can benefit from their manipulation.

The B chromosomes are mainly viewed as self-addicted parasites during their evolutionary life. They are converted into regular elements after stability. It seems that B chromosomes come from a small fragment of chromosomes or genome that has stabilized after fleeing from the control of the cell cycle. They will not be removed during cell division and are transmitted to the next generation.

These genes are important for the initial survival of proto-B. Proto-B contains cell cycle genes and can use cell division to escape elimination. This ability gives the B chromosomes the egoism characteristics to defy Mendel's laws of segregation. As soon as the B chromosome is stabilized, it is free from the strong control mechanism of the cell division and continues to receive new sequences. This mechanism suggests that the B chromosome is like the deposition of genomic variants that can be co-opted by the regular genome at any time. This means that it is no longer just an egoistic element, but it can also contribute to the whole genome. This is probably the reason why the B chromosomes are associated with effects that cover a variety of important biological processes.

History

The idea behind using genetic controls to suppress or alter entire populations, especially those that are considered harmful, by genetic control methods and strategies is not new. It is just the technical skills that scientists have developed recently. As early as the 1940s, scientists such as Serebrovskii and Vanderplank around 1940 and 1944 proposed redirecting an insect's genetic system against itself to destroy insect populations or to make them less destructive (Brandt and Bucking, 2019). Serebrovskii's theory included reducing the ability of insect populations or causing sterile offspring by releasing a large number of mutant

DOI: 10.1201/9781003165316-8

strains, especially strains with chromosomal translocations where there is an exchange of whole segments between different chromosomes. Although the same goal was associated with suppressing or replacing wild populations, none of the examples included changing the evolution of the entire species; there was no drive. However, in the early 1990s, with the advent of genetic engineering, the ability to create artificial genes, recombine different DNA sequences and insert new gene sequences into an organism became more common. The possibility of using different drives for the active propagation of genetic traits in a population seemed to open up.

Chief Traits of B Chromosomes

Innovations are the driver for progress in our insight into hereditary qualities and cytogenetics. Some key inquiries concerning exaggerated B chromosomes include their cause, broad presence, phenotypic impacts, atomic association, methods of legacy and population harmony frequencies. Their fundamental properties can be summed up as follows:

1. They happen in a large number of animal groups; in completely known cases, they are superfluous and people with none are constantly present.
2. They never pair with the standard A chromosomes (A's) of their hosts.
3. Their conduct at meiosis is unpredictable and can prompt disposal.
4. They are generally smaller than the A's and are frequently heterochromatic.
5. In a few cases, they have as of late appeared to convey qualities.
6. They are unsafe to their host life forms when present in high numbers.
7. They have different instruments of amassing, including nondisjunction and meiotic drive. The unique circumstance and foundation of the story are recorded in various audits (Vujošević et al., 2018).

Figure 8.1 is a diagrammatic illustration showing why gene drives are game changers in terms of inheritance, with the ability to maneuver on their own to establish themselves in progeny and subsequent generations (Jones, 2019).

Figure 8.1 shows synthetic gene drives vs. super Mendelian inheritance: in a given group of organisms, a mutation without fitness cost will vanish, in obedience to Mendel's laws, at 50%, while a synthetic gene drive system reaches almost 100%, thus ensuring the proliferation of a trait from generation to generation; this is why gene drives are such game changers.

B chromosomes show variety in size. The morphology of B chromosomes varies from that of A(s) and is for the most part heterochromatinized. In plants, these chromosomes are frequently smaller than the smallest A chromosomes. Even though B chromosomes have been studied to a great extent, nonetheless, data about their atomic piece is sparse.

Target Cells for B Chromosomes

B chromosomes are significantly associated with plants and other eukaryotic cells (Houben, 2017). A target of genetic manipulation by gene drives can be any organism that reproduces itself sexually and the frequency of which is appropriate. Therefore, gene dives are specific for organisms that multiply through a process called meiosis, a special form of cell division that ultimately produces nonidentical sex cells. It is common in eukaryotes or higher organisms but absent in prokaryotes, which excludes all bacteria and archaea (Benetta et al., 2019).

Disadvantages of B Chromosomes

The disadvantages of B chromosomes include:

1. Inefficacy in many organisms.
2. The quick emergence of resistance.

3. Low control, thus irreversibility.
4. The impossibility of containment or recall once released (Brandt and Bucking, 2019).

When resistance is developed against a gene drive, its efficiency can be reduced (Rode et al., 2019).

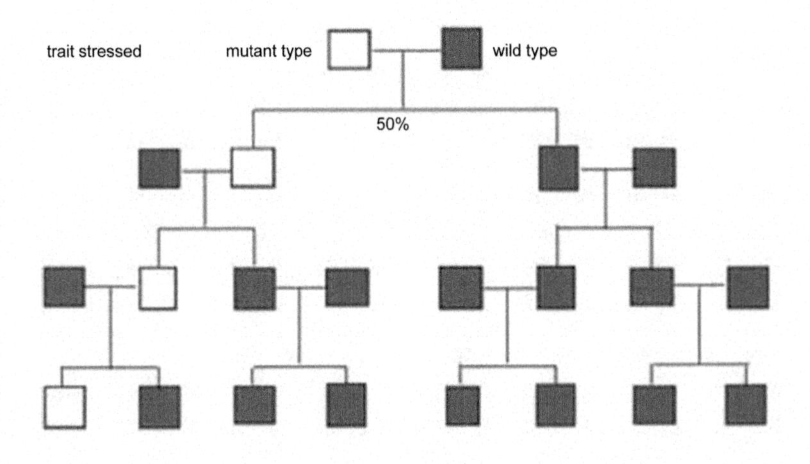

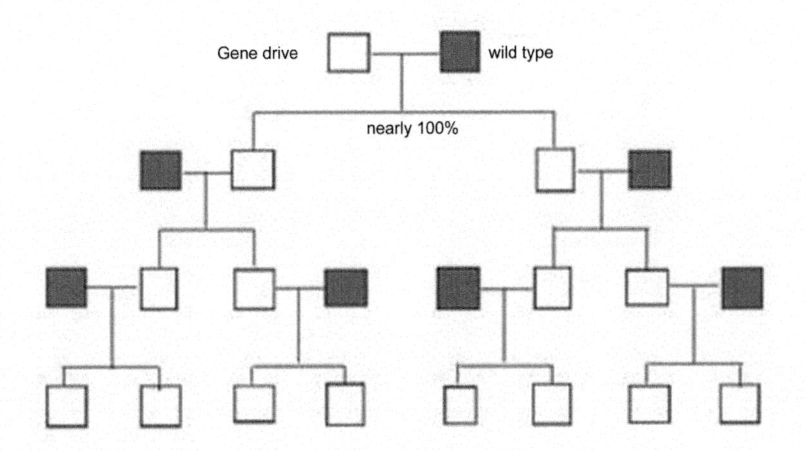

Figure 8.1 Synthetic gene drives vs. super Mendelian inheritance.

(Adapted from Brandt et al., 2019.)

9

Suppression Drives

Mechanism of Suppression Drives

In suppression drives, a genetic modification that induces a decrease in population size is spread. To accomplish this, an altered allele is spread that is harmful to the organism (e.g., causes death or sterility) when two copies are available (homozygous recessive). By replication of the drive in the germline, each offspring remains a heterozygote and as a result is slightly affected by the genetic alteration. This allows genetic modification to spread rapidly through the population initially. Nevertheless, once the modification becomes more common, homozygous offspring will be produced and the population will crash.

Suppression drives rely on the introduction of strongly or mildly deleterious mutations. They are intended to reduce or eliminate a population. The inserted DNA can disrupt a gene, reducing the average fitness of populations (Rode et al., 2019; Johnson, 2018; Sarkar, 2018). When combined with a threshold-independent drive, suppression drives can spread to a global scale and result in the eradication of an entire species. This may particularly become the case if the drive system itself has a low intrinsic fitness cost and no resistance develops to the drive. A population suppression gene drive can be designed to exert a considerable reproductive load on the population as it increases in frequency and sterile homozygous females are produced, thereby limiting reproduction (Harvey-Samuel et al., 2019). Gene drives can be used to suppress vectored diseases through directly depressing vector numbers, possibly to extinction, or suppressing the disease agent during its interaction with the vector (Bull, 2019). The effect of using gene drives in population suppression is shown in Figure 9.1.

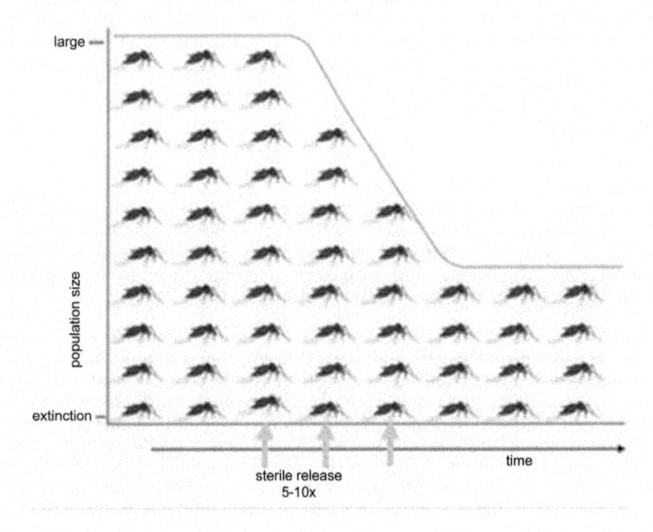

Figure 9.1 An illustration of population suppression by a suppression gene drive. Releasing sterile males into the population can cause transient or permanent population suppression.

DOI: 10.1201/9781003165316-9

Versions

The gene drives that are currently in development or that have already been applied in suppressing populations include meiotic drives (autosomal- or Y-linked X-shredder), maternal-effect dominant embryonic arrest (*Medea*) systems, underdominance-based systems, and homing endonuclease genes (HEG)-based systems, especially CRISPR-Cas. Others, such as killer–rescue, are currently only theoretically explored (Frieÿ et al., 2019; Hoffman et al., 2017; van der Vlugt et al., 2018).

CRISPR gene drives have been widely proposed as a promising potential technology for pest control or even eradication (John and Kean, 2020). This technology provides an ability to disperse genetically engineered or altered genes throughout pest populations with much higher efficiency and prevalence than would be possible via normal genetic inheritance, even with genetic modifications that are deleterious for individuals and populations (McFarlane et al., 2018; Druryetal, 2017).

A Synthetic CRISPR-Cas9 Gene Drive

SgRNA is the guide RNA; Cas9 is an endonuclease that cuts the DNA and Cargo is the desired genetic material added. When all three materials are present in the gene drive cassette, this ensures that each chromosome will have the desired cargo and will be inherited by the next generation, spreading the gene drive.

Construct

CRISPR-Cas9 gene drives fall into the category of HEG drives, first proposed by Burt (2003). They consist of a genetic construct (driver allele) encoding an endonuclease that can cleave a genomic target site and then insert itself into that site. CGD uses a Cas9 endonuclease for cleavage that can be engineered to target virtually any 20 bp-long nucleotide sequence in a genome, using a target-specific guide RNA (gRNA). The driver construct is flanked by two homology arms to facilitate its incorporation at the cleavage site during homology-directed repair (Unkless et al., 2017).

Applications

The fact that gene drive can lead to the spread of fitness-reducing traits (including lethality and sterility) makes it an attractive process to consider exploiting when developing methods to control disease vectors such as malaria-transmitting mosquitoes, ecosystem management and eradicating invasive species and other agricultural pests (Burt and Crisanti, 2018; The Royal Society, 2016; Steinbrecher et al., 2018). Suppression gene drives have been proposed to eliminate *A. gambiae* species to eliminate malaria in Africa (Teem et al., 2019).

Considerable suppression of vector populations can be achieved within a few years of using a female sterility gene drive; although the impact is likely to be heterogeneous in space and time, a driving endonuclease gene targeting female fertility could lead to substantial reductions in malaria vector populations on a regional scale (Yarrington et al., 2018; Marshall et al., 2017; Eckhoff et al., 2017). The exact level of suppression is influenced by additional fitness costs of the transgene, such as the somatic expression of Cas9, and its deposition in sperm or eggs leading to damage to the zygote (Mercier et al., 2019; Godfray et al., 2017). The suppression potential of a drive allele with high fitness costs can be enhanced by engineering where it can also express male bias in the progeny of transgenic males (Beaghton et al., 2019; North et al., 2019; Eckhoff et al., 2017).

Gene drives are also being considered as a way of controlling other invasive species, including wasps in New Zealand (Esvelt and Gemmell, 2017). *Vespula* wasps are haplodiploid, as are many of the other insects listed in the list of 100 of the world's worst invasive insect species.

This includes the red imported fire ant (*Solenopsis invicta*) and sweet potato whitefly (*Bemisia tabaci*) (Sánchez et al., 2020). The haploid males are produced from unfertilized eggs, while females develop from fertilized eggs and are diploid (Unckless et al., 2017). Gene drives have been proposed for use in common wasp management, with spermatogenesis genes as targets (Champer et al., 2019; Dhole et al., 2020). Spermatogenesis genes could be altered in queen wasps to cause viable sperm production in males to fail, resulting in sterile male production (Marshall et al., 2017). Several potential spermatogenesis genes with likely male-specific expression have been identified in wasps. These targets include the *boule* gene, which is essentially for the entry and progression of maturation divisions and sperm differentiation in haploid males. Should a modified queen mate with wild-type males, fertilized worker eggs will be produced, culminating in queen and male production in autumn, all of which will carry the CRISPR cassette and thus propagate it to the next generation and should a modified queen mate with a genetically modified male, fertilization will fail and all offspring in spring will be male (Prowse et al., 2017). This nest will fail and die in spring or early summer, as males do not forage or aid in nest maintenance (Bull et al., 2019).

Examples of Where They Were Used

Much of the work on invasive species has focused on removing rodents from islands (Hu et al., 2019). In these areas, rodents threaten ground-nesting birds. Gene drives are thus being investigated as a potentially more effective and more humane way of eradicating rodents than existing trapping, hunting, or poisoning programs in New Zealand.

Scientists have demonstrated a proof-of-concept population conversion drive that could form the basis of various population suppression strategies in a species of fruit fly whose larvae feed on soft fruits such as cherries. The sterile insect technique eliminated a major cattle pest from almost all of North and Central America, saving billions of US dollars and decreasing the use of insecticides (Edgington et al., 2020; Sumner et al., 2017; Brossarda et al., 2018). The use of "biological" control measures—such as the endosymbiont *Wolbachia*, that when introduced into *A. aegypti* mosquitoes suppresses the transmission of dengue, Zika, and chikungunya viruses—has already seen extensive field trials in Australia (Esvelt and Gemmell, 2017; Kyrou et al., 2018; Wedell et al., 2019).

10

Maternal-Effect Dominant Embryonic Arrest (*Medea*) System

Medea systems are selfish genetic elements that can bias their inheritance by selectively killing non-drive offspring using a toxin–antidote rescue system. They are single-construct designs, meaning all genetic elements involved are tightly linked and transfer as a unit. They were first discovered because of their spread throughout natural populations of the flour beetle *Tribolium castaneum* and the toxin–antidote system was eventually used to create the first synthetic drive system in the fruit fly (Hammond and Galizi, 2017; Buchman et al., 2018). It derives its name from Greek mythology, where *Medea* is said to have killed her children, although it is unknown if it was by accident or actual intent (Esvelt, 2019).

In 2007, researchers from the California Institute of Technology genetically engineered the first gene drive system based on the principles of *Medea* in the model fruit fly *Drosophila melanogaster* using miRNA, *Myd88*, as the toxin to silence an essential embryonic gene. The antidote was the same embryonic gene but modified with an altered sequence so it could not be silenced by the microRNA (Hammond and Galizi, 2017; Burt and Crisanti, 2018; Esvelt, 2019).

According to the models, *Medea* is regarded as a strong drive system that could spread payload genes rapidly if it is released at high frequencies and the fitness cost is kept low. Recent laboratory experiments carried out with the spotted wing *Drosophila* (*D. suzukii*), an agricultural pest in soft fruit production in California, confirmed the need for high release frequencies. They also showed in long-term cage trials that selection for resistance to the miRNA-based toxin being used is a concern (Min et al., 2018; Buchman et al., 2018).

Since then, several similar strategies have been proposed that vary in the arrangement and mechanism of killer-rescue components, as well as the release threshold predicted for these to drive population replacement or population suppression. The female makes a toxin during oogenesis that will lead to the death of the embryos unless any of them has inherited a copy of the *Medea* element from its mother or father, as this also holds the antidote within the same element (Backus and Delborne, 2019).

There is a multitude of different theoretical systems inspired by these *Medea* principles, the closest of which are Merea, inverse *Medea*, and *Semele*, used for modeling of engineered gene drives in order to see, for example, whether gene drives would be less invasive or suppression could lead to population collapse if payload genes would easily find fixation in a population (Esvelt, 2019). The long-term dynamics of a threshold-dependent *Medea* drive in a two-population system is shown in Figure 10.1.

The goal is to drive the *Medea* drive to fixation in one population without spreading to fixation in a neighboring population. In Figure 10.1: (a) There are two populations, initially consisting of only wild-type individuals, and there is some bidirectional migration between the populations. Males that are homozygous for *Medea* are released into population 1 so that 55% of the males in population 1 will be carrying the *Medea* drive (roughly 1.22:1 ratio). Over several generations, these populations mate and migrate. (b) After 200 generations, the fate of the two-population system depends on the migration rate and the relative fitness of the gene drive. two separate *Medea* drives can be released into one population, with an expected migration rate of 0.2% per generation. Ellipses around either point represent parameter uncertainty in both gene drive fitness and migration rates. Because gene drives are more likely to impose an ecological fitness cost on the organisms carrying them, there is a higher chance that ecological fitness would be lower than expected, with only a small chance that ecological fitness is higher than expected. The triangle is a drive that was developed such that it has an expected relative fitness of 75% of wild-type counterparts. If the ecological dynamics of the system lie anywhere

DOI: 10.1201/9781003165316-10

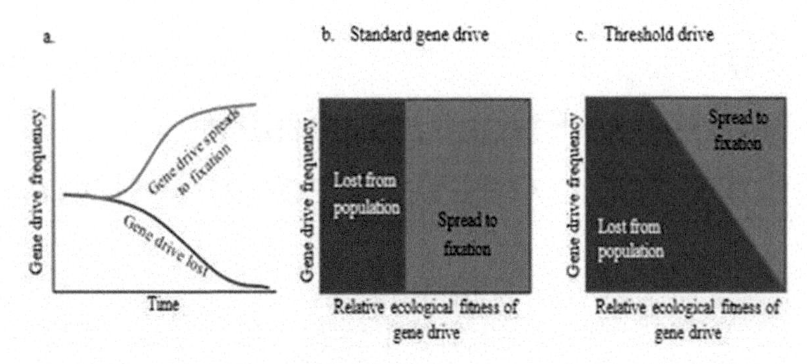

(Adapted from Backus and Delborne, 2019)

Figure 10.1 Example of long-term dynamics of a threshold-dependent *Medea* drive in a two-population system.

in the region of uncertainty, the gene drive will go to fixation in only the target population. The other drive is a drive developed with a higher expected relative fitness of 90%. In this case, there is a small probability that *Medea* could drive to fixation in both populations (Backus and Delborne, 2019).

In standard and threshold gene drives, two general types of long-term dynamics depend on the frequency and fitness of the gene drive. All examples here are conceptual, not relating directly to any particular gene drive system. Most gene drives will either spread to fixation or be lost from a population over time. Regardless of frequency, standard gene drives spread to fixation with high relative fitness and are lost from a population with low relative fitness. The long-term dynamics of threshold drives depend on both gene drive frequency and gene drive fitness. A threshold gene drive is more likely to spread with higher frequencies and lower fitness.

11 Heritable Microorganisms: *Wolbachia*

Wolbachia are obligate symbionts that live in vacuoles inside the cells of their host. Infections are passed on vertically in egg cytoplasm, from infected mothers to their offspring. Transmission by infected males is barred by the low volume of cytoplasm in mature sperm, though rare cases may happen. Since *Wolbachia* are maternally transmitted, infections only can spread in the host population if infections favor the creation of offspring by infected females. *Wolbachia* induce four well-known types of effects on their hosts to accomplish this goal, that is, feminization, parthenogenesis, male-killing, and cytoplasmic incompatibility (CI).

Wolbachia are one of the notable pandemics of life from a biodiversity standpoint because they are estimated to occur in millions of invertebrate species, together with 40% of all arthropod species (Taylor et al., 2018). They developed a collection of host reproductive manipulations that proliferated their worldwide dominance in arthropods, comprising feminization, parthenogenesis, male-killing, meiotic drive and CI.

Wolbachia are obligate endosymbiotic proteobacteria that are most closely linked to the genera *Ehrlichia* and *Anaplasma*. They are pleiomorphic, stretching from 0.2 to 4μm in size and exist in an obligate intracellular niche inside host-derived vacuoles. The cell is gram-negative, with an inner and outer membrane, but has lost much of the characteristic cell wall structure, as well as peptidoglycan (Taylor et al., 2018). Extracellular transfer happens momentarily in nematodes but is limited to the transfer from hypodermis to ovary within the host. In arthropods, transfer between different hosts happens repeatedly and even plants can act as a transitory bridging "hosts". Because *Wolbachia* are mostly transmitted vertically through the host maternal germline, they have an extraordinary capability to adjust core host cellular and developmental processes to favor infected females. This consists of changing the host's chromosome condensation and cell cycle timing, sex determination pathways, apoptosis, stem cell biology and axis determination during early embryogenesis.

In terms of evolutionary theory, CI can be assumed in the light of intragenomic conflicts that may arise from the maternal transmission of the bacteria. As males are reproductive dead ends for cytoplasmic elements, it is beneficial to "utilize" them to increase transmission through female hosts. CI is one of four strategies of such "reproductive parasitism": males are exploited by the bacteria to kill the offspring of uninfected females so that infected females have a selective advantage. The other three phenotypes of reproductive parasitism (male-killing, feminization, and parthenogenesis induction) all involve sex-ratio distortion. CI is commonly acknowledged within a modification-rescue (or poison-antidote) basis: sperm is modified by the bacteria in males, and the same or a related strain must be existing in the eggs to "rescue" the modification, allowing the progeny to develop normally.

The exact nature of this modification–rescue mechanism is currently not understood. Cytological studies have presented that in incompatible crosses, the paternal chromosomes do not condense, are damaged and ultimately are lost during the first mitotic divisions. In diplodiploid and most haplodiploid species, the loss of paternal chromosomes leads to disordered development and death of the embryos.

History

Wolbachia is a genus of obligate gram-negative αα-proteobacteria. The bacteria infect cells of arthropods, 60–70% of insects and also some nematodes (Dionysopoulou et al., 2020; Pan

DOI: 10.1201/9781003165316-11

et al., 2018; Sicard et al., 2019). The intracellular endosymbiont is the most common parasitic microbe and is possibly the most common reproductive parasite in the biosphere (Caragata and Moreira, 2017; Kajtoch and Kotásková, 2018; Schebeck et al., 2019).

The bacteria genus was first identified in 1924 through collaborative work between Marshall Hertig, an entomologist and Simeon Burt Wolbach, a pathologist, in their study of the common house mosquito (*Culex pipiens*) (Caragata and Moreira, 2017). The study was to be published later by Hertig in 1936 and described the species as *Wolbachia pipientis*.

Marshall (1938) was the first to notice that crosses between certain strains of the mosquito *Culex pipiens* were incompatible in one direction (that is, males from strain *A* and females from strain *B*), whereas the opposite direction (*B* males × *A* females) produced viable progeny. This process was further investigated in the 1940s and 1950s. These early investigations showed that the incompatibility trait was inherited maternally, declaring that to be an extranuclear causative agent and giving rise to the term "cytoplasmic incompatibility" (CI). In addition to unidirectional CI, bidirectional CI was also observed between strains of *C. pipiens*. During the early 1970s, a link was drawn up between CI and the intracellular α-proteobacterium *Wolbachia* that had long been noted to infect *C. pipiens* (Beckmann, Sharma et al., 2019). Unidirectional CI commonly occurs when infected males mate with uninfected females but can also be found in crosses between individuals infected with different strains of bacteria. In bidirectional CI, both directions of a cross are incompatible due to infection with different strains of CI-inducing microorganisms. In both cases, offspring from incompatible crosses go through mortality at the early stages of their development. Uni- and bidirectional CI are illustrated in Figure 11.1.

The two graphs in Figure 11.1 display success (tick marks) or failure (crosses) of progeny production of crosses between parents with different infection states. Blank symbols in the parent generation mean that these parents are uninfected, although the two shades of gray signify infection with two different strains of bacteria. With unidirectional CI, only crosses between infected males and uninfected females are incompatible. With bidirectional CI, crosses between males and females infected with different strains of CI-inducing bacteria are incompatible. Note that sometimes unidirectional CI is also detected between hosts infected with different strains of bacteria.

The *Wolbachia* bacteria species are taxonomically classified under the domain Bacteria, phylum Proteobacteria, class Alphaproteobacteria, order Rickettsiales, family Anaplasmataceae, tribe Wolbachieae, and genus *Wolbachia* (Chrostek and Gerth, 2019; Leggewie et al., 2018; Taylor et al., 2018). The bacteria species are usually referred to by their genus without identifying distinct species. However, they belong to about 17 phylogenetic clades called supergroups (A to Q) found in arthropod and nematode hosts (Bridgeman et al., 2018; Carvajal et al., 2019; Kajtoch and Kotásková, 2018). Distinct *Wolbachia* strains of each supergroup are named according to their host species (e.g., *w*Pip in *Culex pipiens*, *w*Mel in *Drosophila melanogaster*, *w*Alb in *Aedes albopictus*, *w*Ri in *D. Simulans*, *w*Ana in *D. ananassae*, *w*Ano in *D. anomalata*, and *w*Pan in *D. pandora* (Chouin-Carneiro et al., 2020; Fraser et al., 2017; Zhang et al., 2016). If significant genetic polymorphism is revealed by studies within an already defined "strain", the new information, including sampling location or phylogenetic position, can be added to name the new strains to uniquely separate them (Sicard et al., 2019).

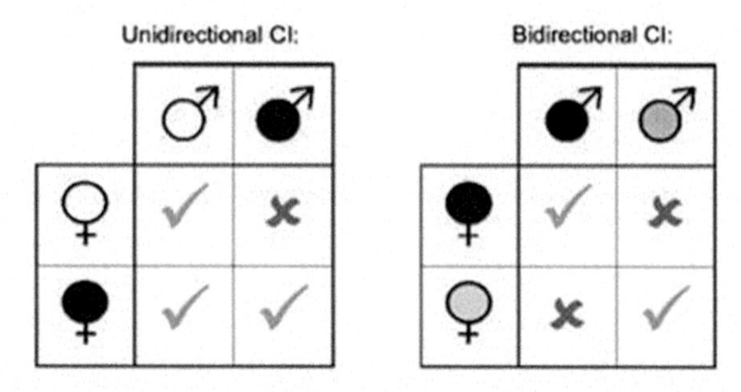

Figure 11.1 Illustration of uni- and bidirectional cytoplasmic incompatibility (CI).

Since the 1970s, many scientists have devoted considerable effort and time to the study of this group of bacteria. These studies have been motivated by the need to deal with vector transmission of diseases that have affected human and animal health, its ubiquitous distribution, its many different evolutionary interactions and its potential use as a biocontrol agent. *Wolbachia* species have been shown to manipulate many aspects of the biology of their hosts in different ways (Kodandaramaiah, 2008; Ross et al., 2019; Sicard et al., 2019).

Target Cells

The work of Hertig and Wolbach on *Culex pipiens* revealed granular, circular bacterium in mosquito eggs, larvae and adults, with the greatest bacterial density observed in the gonads (Bi and Wang, 2019; Caragata and Moreira, 2017).

Subsequent research has revealed that *Wolbachia* bacteria infect host reproductive tissues, especially the ovaries, where the bacteria are ubiquitous in individual oocytes and nearby nurse cells. This is the basis of vertical transmission from mother to progeny through the cytoplasm of eggs (Caragata and Moreira, 2017; Flores and O'Neill, 2018; Sicard et al., 2019). In males, *Wolbachia* is found in the germline cells of the testes, but mature spermatozoa lack *Wolbachia* (Lindsey et al., 2018). Some of the vector insects whose cells can be infected by *Wolbachia* include major vector mosquito species of *Aedes*, *Anopheles*, and *Culex* genera (Maciel-de-freitas et al., 2012; Moyes et al., 2019; Sicard et al., 2019). Other vectors that have also been shown to have *Wolbachia* include bed bugs, tsetse flies, blue butterflies and fruit flies, among others. There is evidence of the presence of *Wolbachia* in somatic cells of different tissues in some insect species. Its presence has been attributed to causing effects on behavior and immunity against pathogens (Bi and Wang, 2019; Lei et al., 2020a; Pan et al., 2018).

Ease of Use

The use of *Wolbachia* for sterile insect technique (SIT) and gene drives is fairly simple for producing transgenic vectors and noninvasive populations. The SIT and gene drive strategies take advantage of CI to suppress or eliminate vector populations or replace them with ones showing reduced fitness or decreased vector competence without the use of complex genetic engineering and irradiation (Moretti et al., 2018a).

Specificity

While different insect species are naturally infected with specific strains of *Wolbachia*, other species are naturally uninfected. For *Culex pipiens* and *Aedes albopictus*, almost all species harbor *Wolbachia*; *Anopheles* species and *Aedes aegypti* have been generally found to be uninfected (Sicard et al., 2019). *Anopheles* mosquitoes, the major vectors of *Plasmodium*, were considered to be exempt from *Wolbachia* because classic PCR diagnostic tests were always found negative. However, deep sequencing of *Wolbachia*-specific 16S rRNA recently suggested putative natural infections of *Anopheles coluzzii* and *Anopheles gambiae* in Burkina Faso. The *Wolbachia* 16S sequences obtained were attributed to a new strain named "*wAnga*" (Sicard et al., 2019).

CI studies have shown that the artificial transinfection of heterologous *Wolbachia* strains into different insect species usually still causes CI. The phenotypic consistency across species suggests that *Wolbachia*-induced CI targets conserved cellular machinery required for cell and nuclear division and hence is not species specific (Caragata and Moreira, 2017; Jeffries et al., 2018).

Error Rate and CI Efficiency

The error rate and efficiency of *Wolbachia* gene drives depend on whether we always get the intended outcome of CI. In other words, the error rate is a measure of the predictability of the CI outcome. Naturally, CI is strong in some species and weak in others. The parasitic wasp *Nasonia vitripennis* is one of the arthropods that naturally show high CI, in the range of 90%. The beetle *Chelymorpha alternans* has CI of about 90% when doubly infected with *wCalt1* and *wCalt2*. Double-infected males are incompatible with single-infected (*wCalt1*) females (Funkhouser-Jones et al., 2018).

The decrease of the prevalence of CI with successive mating exhibited in infected males implies a low error rate and high efficiency for young males and a high error rate and low efficiency for older males. In other words, while the error rate is directly proportional to the age of infected male mosquitoes, efficiency is indirectly proportional to the same. For female hosts that exhibit bidirectional CI, there are low error rates and high efficiency to drive a change into a population (Macias et al., 2017; Sicard et al., 2019).

Principle

Studies around *Wolbachia* have been centered on their ability to manipulate and modify the physiology of their host, especially reproduction. Yen and Barr discovered that *Culex* mosquito embryos in eggs were killed by CI when the sperm of *Wolbachia*-infected males fertilized infection-free eggs (Sicard et al., 2019).

Wolbachia bacteria enhance their spread by maternal vertical transmission by altering the reproductive capabilities of their hosts in four different ways:

Male killing (MK). It occurs when infected males die during larval development, increasing the proportion of infected females in the population (Asimakis et al., 2019; Kajtoch and Kotásková, 2018; Perlmutter et al., 2019; Reynolds et al., 2019). MK *Wolbachia* are one of the numerous types of maternally inherited bacteria that kill males throughout embryonic development. Henceforth, infected females could lay a mixed brood of male and female eggs, but only the later survive to develop into adults. Cases of MK *Wolbachia* have been described in taxa including *Coleoptera* (*Coccinellidae, Tenebrionidae*), *Diptera* (*Drosophilidae*), and *Lepidoptera* (*Nymphalidae*). The cytological mechanisms initiating male killing are unidentified (Harumoto et al., 2018).

Feminization. Infected males develop into females or infertile pseudo females in a process called feminization as *Wolbachia* results in the elimination of paternal chromosomes thereby converting a parasitic wasp (*Nasonia vitripennis*) diploid embryo into a haploid male (Asimakis et al., 2019; Chegeni and Fakhar, 2019). Sex in this and other crustaceans is determined by the action of a male hormone that suppresses female development. *Wolbachia* is believed to inhibit the development of the androgenic gland that makes this hormone and also may block receptor sites vital for hormone activity (Kageyama et al., 2017). Therefore, infected isopods produce female-biased sex-ratios irrespective of their sex chromosome complement (WZ?/females; ZZ?/males).

Parthenogenesis induction (PI). Infected females reproduce without mating with males, in a process of parthenogenesis (Asimakis et al., 2019; Chegeni and Fakhar, 2019; Kajtoch and Kotásková, 2018). PI *Wolbachia* only are known from haplodiploid taxa. Reproduction by arrhenotoky is popular in these taxa, whereby males grow from unfertilized (haploid) eggs and females grow from fertilized (diploid) eggs. Two types of "gamete duplication" describe cases of *Wolbachia*-induced parthenogenesis in *Hymenoptera*. In *Leptopilina clavipes* (*Figitidae*) and *Trichogramma* spp. (*Trichogrammatidae*), infections obstruct the separation of chromosomes in anaphase of the first mitotic division. The unfertilized egg therefore contains a diploid nucleus with two identical sets of chromosomes and matures into a female. In *Muscidifurax uniraptor* (*Pteromalidae*) and *Diplolepis rosae* (*Cynipidae*), the first mitotic division is normal and yields haploid nuclei, but these afterward fuse to restore the diploid condition. Females rising from gamete duplication are homozygous for all alleles (Ma and Schwander, 2017).

Cytoplasmic incompatibility (CI). This is the utmost common of the *Wolbachia*-induced phenotypes that affect host reproduction. It has been described in *Coleoptera, Diptera, Homoptera, Hymenoptera, Lepidoptera, Orthoptera*, mites (*Acari: Phytoseiidae, Tetranychidae*) and woodlice (*Isopoda*: *Porcellionidae*). CI rises in matings between *Wolbachia*-infected males against uninfected females or between partners infected with dissimilar strains of *Wolbachia*. The existence of *Wolbachia* in males is understood to introduce a factor into their sperm that inhibits embryogenesis in the fertilized egg, except the female partner is infected with a similar *Wolbachia* strain to permit the sperm's "rescue". Unidirectional CI is most common and frequently happens between males infected with a single strain of *Wolbachia* and uninfected females. CI makes *Wolbachia*-infected males unable to reproduce with uninfected females or females infected with a different and incompatible *Wolbachia* strain (Beckmann, Bonneau et al., 2019; Kittayapong et al., 2018; Shropshire and Bordenstein, 2019; Ün et al., 2019). CI causes the damage of paternal chromosomes with different results depending upon whether the host is a diplodiploid or haplodiploid species. In diplodiploid species, paternal chromosome damage causes the haploidization of fertilized eggs that successively die. When a similar process occurs in haplodiploid species, theoretically, all of the haploid eggs mature into males, that is, male-development (MD) type CI. Then again, CI in haplodiploid species may be categorized by the death of embryos intended to become females, that is, female-mortality (FM) type CI. Both MD-type and FM-type CI produce male-biased progenies of offspring. Nevertheless, they can be distinguished because the latter also results in fewer offspring relative to the amount produced in compatible crosses. Whereas MD-type CI most likely mirrors the thorough loss of paternal chromosomes, FM-type CI may be due to their incomplete loss, initiating lethality (Beckmann et al., 2017).

Cytoplasmic Incompatibility (CI)

CI in its simplest form occurs when eggs are laid as an outcome of fertilization between an uninfected female and infected male and are rendered nonviable. However, *Wolbachia*-infected females lay viable eggs irrespective of the infection status of the male and may transmit *Wolbachia* to their offspring (Mishra et al., 2018). CI is expressed as embryonic lethality in crosses between infected males and uninfected females. This lethality is completely rescued by females infected with the same or a similar *Wolbachia* strain (toxin–antidote system). CI is also regarded as a form of conditional infertility, where fertilization does not successfully result in embryo development when sperm from a *Wolbachia*-infected male fuses with eggs from a female that is not infected or one that is infected with a different and incompatible *Wolbachia* strain (Beckmann, Sharma et al., 2019; Chen et al., 2019).

Two events lead to CI. These are a modification of sperm by *Wolbachia* and rescue of the modified sperm upon fertilization. Following fertilization, the modified sperm may successfully develop into an embryo if the rescue factor is present in the egg, or embryogenesis will be unsuccessful if the egg lacks a relevant rescue factor that is *Wolbachia* strain-specific (Beckmann et al., 2017; Frederick et al., 2019). Studies have revealed that *Wolbachia* is absent in mature sperm but is found in the germline tissue of the testes, spermatocytes, and some spermatids. It has also been shown that *Wolbachia* are required in spermatocytes and spermatids for the modification of sperm. During maturation of spermatids, they change shape so that they can easily swim through the removal of cytoplasm, development of tail and concentration of mitochondria. Excess fluid from the cytoplasm including *Wolbachia* gets squeezed out in the process. It is this process that ensures mature spermatozoa are free of *Wolbachia*.

Since CI weakens in males with each successive fertilization of eggs, it means that *Wolbachia* stockpiles a limiting factor that induces CI in the mature spermatozoa. The strength of CI, in this case, is evaluated by the percentage of embryos that become nonviable. With continued spermatogenesis, the limiting factor gets depleted. The strength of CI in males is proportional to the quantity of *Wolbachia* cysts in spermatocytes (Beckmann et al., 2017).

Unidirectional and Bidirectional CI

When a modified sperm from an infected male fertilizes an egg from an uninfected female, CI manifests as unidirectional CI. However, with an infected female, the CI is bidirectional. Bidirectional incompatibility occurs when the *Wolbachia* strain in the male is incompatible with the strain in the female. This form of CI has revealed the mosquito species harbor a large diversity of cross-incompatible *Wolbachia* strains (Ant et al., 2020; Beckmann, Sharma et al., 2019; Moretti et al., 2018b).

Wolbachia strains evolve differently in different hosts, resulting in CI systems that are bidirectionally incompatible. Such evolutionary developments may be attributed to *wPip* from *C. pipiens* have two related operons. The second pair of paralogous genes were named cinA (WPA0294) and cinB (WPA0295). Characteristics of CidB and CinB show that they are potentially homologous, but cidB lacks the putative nuclease active site domain (Beckmann et al., 2017).

Wolbachia Genes Inducing CI

Cytological studies of mature *C. pipiens* sperm cells suggested that a toxin deposited in sperm by *Wolbachia* is behind CI. This hypothesis led to the discovery of a protein coded by the *Wolbachia* gene, WPA0282. The gene is part of a two-gene polycistronic operon, where the second gene was identified as WPA0283 coding for the CI-inducing deubiquitylating enzyme (DUB). The two genes were renamed cidA (wPa_0282) and cidB (wPa_0283). The function of cidA was suggested to be an antidote to rescue the viability of modified sperm from infected males, since its expression during early oogenesis restores sperm viability. Further studies suggest that cidB could also act as a toxin (Beckmann et al., 2017; Beckmann et al., 2019).

Mechanism of CI

Although precise mechanisms are still vague, the molecular mechanism behind CI is modeled as a modification–rescue (or toxin–antidote) system whereby a sperm that undergoes *Wolbachia*-mediate modification can be rescued in the egg by a *Wolbachia*-encoded factor. After fertilization, the nuclear envelope of the sperm pronucleus breaks down and exchanges protamines with maternal histones. Protamines are small basic proteins used to package paternal DNA at high density in the spermatozoa. Then the maternal and paternal pronuclei mix and undergo first zygotic mitosis. Chromosomes from both pronuclei condense at the same time, align at metaphase, and separate during anaphase. In cases that exhibit CI, the paternal chromatin fails to condense properly because of impaired maternal histone deposition on paternal DNA. This results in lethal and abnormal segregation and bridging of paternal DNA at anaphase. This cellular phenotype at first zygotic mitosis is the principal cause of CI.

Toxicity takes place if cells can no longer synthesize the proteins because the antidote protein degrades much more rapidly than the toxin, thereby releasing active toxin. It is assumed that CidA and CidB proteins behave similarly. The toxicity of cidB may be attributed to its deubiquitylating and subsequent loss of function of histone chaperones and H2 histone, which are both essential for the formation of nucleosomes (Beckmann et al., 2017; Beckmann, Sharma et al., 2019).

Two wide-ranging biochemical models have been suggested: either (a) *Wolbachia* in the male produce a product that interrupts sperm processing in the egg (except rescued) or (b) bacteria in the male act as a "sink" to bind away a product required for normal processing of the sperm in the egg (Asselin et al., 2019). The biochemical mechanisms of CI continue to be unfamiliar, and this undoubtedly is a major area for research. In recent times, consistent with the "sink" hypothesis, several host chromatin-binding proteins (such as H1 histone-like protein) have been noted to bind to *Wolbachia* inside host cells. Also, *Wolbachia* can now be kept in insect cell cultures, which would enable biochemical and genetic studies.

Cytogenetic mechanisms of CI have been studied in *C. pipiens*, *D. simulans*, and *Nasonia vitripennis*. In all cases, CI is linked with early mitotic faults in the fertilized egg. In the parasitic wasp *N. vitripennis*, it has been revealed by a combination of genetic and cytogenetic studies that the paternal chromosomes form a diffuse chromatin mass in the first mitosis, fail to go through segregation, and usually are lost in later divisions. Paternal genome loss results in the development of haploid (male) progeny in entities with haplodiploid sex determination, whereas a comparable mechanism in diploid species would end in embryonic death.

Fragmentation of the chromatin mass also can happen, with paternal chromatin segregating to some daughter nuclei. Paternal chromosomes with large terminal deletions sometimes survive and can be transferred to the next generation. In diploid species, for instance *D. simulans*, both irregular first mitosis and later stage disruptions in embryogenesis have been detected. Lassy and Karr found that the paternal pronucleus and its related chromosomes display abnormalities in a large percentage of CI-expressing embryos. The later stage abnormalities detected in embryos of *Drosophila* spp. could be because of CI-induced aneuploidy, as has been suggested for higher mortality levels observed when males with antibiotically lowered bacterial densities are used in CI crosses in *Nasonia* spp. CI in these very different organisms rather possibly has a mutual basis, involving disruption of one or more processing steps of the paternal pronucleus after fertilization. Yet this remains to be determined.

Many factors have been suggested to influence the expression of CI, including bacterial strain, host genotype and bacterial density. These factors can cooperate in multifaceted ways to influence the strength and direction of CI. The data is clear that bacterial strain plays a significant role. Further evidence comes from studies of bidirectional incompatibility. In every single case so far studied, bidirectional incompatibility happens between different strains of *Wolbachia*, centered on sequence differences in 16S rDNA and/or ftsZ. In *D. simulans*, three diverse naturally occurring bacterial strains have been recognized that are bidirectionally incompatible with each other. All three are A-*Wolbachia* strains. The pattern shows that new compatibility forms can evolve equally quickly within a bacterial group. Less remarkably, bidirectional incompatibility has been revealed between A-*Wolbachia* and B-*Wolbachia* found in *Nasonia* wasps and between the *D. simulans* Riverside (R) strain (an A-*Wolbachia*) and an *A. albopictus* B-*Wolbachia* strain initiated in *D. simulans* by microinjection. We at present cannot distinguish whether there is a multitude of bidirectional compatibility types or a relatively limited set.

Procedures

Aedes aegypti mosquito will be used for describing experimental procedures that are done with *Wolbachia*-infected insects.

Collection of Mosquito Stock

Wild-type mosquitoes are collected periodically using ovitraps (with flannel cloth strips for oviposition) placed at different locations. Ovitraps are set for 1 week, and strips are then returned to the laboratory, kept damp for 2 days and then air-dried for 1 day. Eggs are hatched, reared to the 3rd–4th instar larvae and *A. aegypti* are retained and fed sheep blood with a Hemotek membrane feeder or hamsters. Resulting eggs (F1) are harvested weekly for 4–8 cycles. Eggs are kept damp for 2 days, dried for 1 day and stored in a humidified sealed plastic container and kept at room temperature (25°C ± 2°). Ovitraps may be designed as plastic buckets 13 cm in diameter and 12 cm high. Each bucket is filled with approximately 750 ml of water, with 6 pellets of rabbit food as bait to attract females. Two cotton flannel strips are added to the sides of the bucket to support oviposition (Tantowijoyo et al., 2020).

Sterilizing *Wolbachia* from Infected Mosquitoes

The uninfected stock of mosquitoes is created by treatment with antibiotics to create negative control stocks. Tetracycline and rifampicin are examples of antibiotics that can be used for such treatment. The treatment is repeated over at least two generations in experimental cages. PCR amplification of 952 bp 16S rDNA gene fragment and the 560 bp wsp gene fragment (Hu et al., 2020) using strain-specific *Wolbachia* primers should be used to validate the negative infection status of the treated stock. The antibiotics may be administered by addition to the water used in growing larvae and pupae (Reveillaud et al., 2019). Sterilization can also be achieved by radiation for those species that are tolerant to it. However, for most mosquitoes species, radiation weakens them and the experimental outcome gets compromised (Kittayapong et al., 2018).

Infecting Experimental Mosquitoes with *Wolbachia*

Wolbachia can be experimentally transfected between species by microinjection into eggs, for instance, from *Aedes albopictus* to *Aedes aegypti* (O'Neill et al., 2018).

Sexing

Wolbachia-infected mosquito studies rely on the effectiveness of separating males and females. Many insect species are characterized by sexual size dimorphism. In mosquitoes, female pupae are usually larger than their male counterparts. This trait can be efficiently used for sexing mosquitoes using mechanical separation means such as standard sieves, the Fay-Morlan glass plate (Kittayapong et al., 2018) and the McCray adjustable opening separation system. These tools have proved reliable in sexing *Aedes aegypti*, *Aedes albopictus*, and *Culex quinquefasciatus* species. However, there is insignificant sexual size dimorphism in *Anopheles albimanus* and *Anopheles arabiensis*; hence, these sex separation tools are ineffective. For species that lack significant sexual size dimorphism, male mosquitoes may be selected by the addition of toxicants to blood meal to eliminate females. Another method involves using genetically modified mosquitoes with a sex-linked marker (Mashatola et al., 2018; Ross et al., 2017; Zheng et al., 2019).

Harvesting Sperm

Wolbachia deposit CI- inducing factors in the spermatozoa. For experiments to analyze the nature of these proteins, sperm is conveniently harvested from the spermathecae of the female mosquito. The sperm may present in any one of two sperm storage organs: the paired spermathecae or a single seminal receptacle. The spermathecae stores sperm in the female after mating and releases it later to fertilize mature eggs. The insect is dissected and the reproductive organs are removed with the help of a microscope (Avanesyan et al., 2017).

Analysis of Limiting Factor

The limiting factor protein in the sperm should be coded in the genome of *Wolbachia* and not the host. The protein should be separated and analyzed by either liquid chromatography, mass spectrometry, SDS-PAGE, or their combination. Further bioinformatics analyses are carried out using protein sequence data by BLAST-related tools to identify homologous gene sequences in *Wolbachia* (Beckmann et al., 2017).

Applications

The ability of *Wolbachia* to induce CI is pivotal in implementing the two major *Wolbachia*-based vector control strategies: to reduce the population size of *Wolbachia* host vectors or decrease population density in a locality by killing individuals and arresting reproduction processes and to change the population with undesirable characteristics such as spreading diseases by gradually replacing with individuals without such negative traits. These two strategies are categorized in the following sections.

Sterile Insect Technique

The first method, called the "incompatible insect technique" (IIT), is related to the classical sterile insect techniques (SIT) and aims at decreasing mosquito population size by releasing *Wolbachia* infected males that are incompatible with local females. In this strategy, the local females produce nonviable embryos, resulting locally and temporally in the vector population crash-down (Beckmann, Sharma et al., 2019; Sicard et al., 2019).

Gene Drive

The second method uses CI induction by *Wolbachia* not to reduce the density of a focal vector population but to sustainably replace its uninfected individuals by *Wolbachia* infected ones. Indeed, it has been shown that *Wolbachia* can interfere negatively with the transmission of disease pathogens, including the major arboviruses chikungunya, dengue, Rift Valley, West Nile, Zika, and so on. In this strategy, CI allows the progressive invasion of the local vector population with individuals harboring *Wolbachia*, which mediates the blocking of arboviruses transmission (Beckmann, Sharma et al., 2019; Sicard et al., 2019). CI can be used to spread desirable genes found in the *Wolbachia*-infected strain by reducing the survival rate of individuals with deleterious genes that are in the wild-type host population (Macias et al., 2017).

Wolbachia Causing Disease

Besides insects, *Wolbachia* infect a range of isopod species, spiders, mites and numerous species of filarial nematodes (a kind of parasitic worm), as well as those causing onchocerciasis (river blindness) and elephantiasis in humans, plus heartworms in dogs. Not only are these disease-causing filarial worms infected with *Wolbachia*, but *Wolbachia* also appears to play an excessive role in these diseases. A huge part of the pathogenicity of filarial nematodes is owed to host immune reaction to *Wolbachia*. Eradication of *Wolbachia* from filarial nematodes usually results in either death or sterility of the nematode. As a result, current approaches for control of filarial nematode diseases consist of elimination of their symbiotic *Wolbachia* using the simple doxycycline antibiotic instead of directly killing the nematode with far more toxic antinematode medications.

Advantages

Knowledge of vector microbiota has been used to develop new vector control strategies that offer environmentally friendly ways to control insect pests and disease vectors by avoiding the use of polluting insecticides (Lei et al., 2020b; Mishra et al., 2018; Sicard et al., 2019). *Wolbachia*-driven CI provides an alternative solution in pest populations that have developed pesticide resistance. The technology also proffers organic, safe and less risky solutions to problems as compared to the use of genetic engineering. *Wolbachia* gene drives have very little effect on ecosystems, as there is no release of any substance that may affect other unintended species. *Wolbachia* gene drives are fairly easy to design and implement, as they can rely on naturally infected host strains (Neve, 2018; Sicard et al., 2019). *Wolbachia*-mediated control initiatives require infected individuals to be released and spread at high frequency before genes can self-sustainably spread. This makes them safe drive mechanisms to solving vector and pest population problems that are localized and that may need to be reversed (Leftwich et al., 2018).

Disadvantages

In some cases, *Wolbachia* enhance the development of diseases caused by their hosts. It has been shown that there is a significant contribution by *Wolbachia* to the pathogenicity of host filarial nematodes, as elimination of *Wolbachia* from the nematodes makes them sterile and may even be fatal. Hence, control of filarial nematode diseases involves disruption of the symbiotic relationship by administering simple doxycycline antibiotics rather than medication that directly kills nematodes (Chegeni and Fakhar, 2019; Newton and Slatko, 2019).

Regardless of their successes, there are limitations of *Wolbachia* infections that may disturb their efficacy as disease control agents. The bulk of *Wolbachia* infections in *A. aegypti* decrease mosquito fitness and these costs tend to be worsened in stressful environments such as when larvae are starved or in quiescent eggs. Fitness costs can have huge effects on invasion accomplishment. For instance, the *w*MelPop infection failed to persevere in release zones in Australia and Vietnam regardless of reaching frequencies above 90%, likely due to the substantial fitness costs of this strain (Fraser et al., 2017). *Wolbachia* infections that happen naturally in mosquitoes can obstruct patterns of CI and limit the prospective for population replacement and suppression. Density-dependent interactions and spatially heterogeneous environments can also slow the rate of invasion, as can pesticide susceptibility in released mosquitoes.

For population replacement programs to be effective, *Wolbachia* infections should continue at high frequencies in the environment and block virus transmission under field conditions for many years after deployment. There is a risk that *Wolbachia* infections, viruses, or mosquitoes will evolve following the setting up of *Wolbachia* in populations, leading to a reduced amount of effective virus protection in *Wolbachia*-infected mosquitoes in the long term. Yet, the *w*Mel infection has stayed steady so far in terms of virus blockage and its effects on fitness. After seven years in the field, *w*Mel has reserved a high titer and remains to induce complete CI in the laboratory, showing that attenuation is not likely for at least several years following deployment.

Wolbachia strains in *A. aegypti* differ in their response to heat stress; the *w*Mel and *w*MelPop strains are relatively susceptible, while *w*AlbA, *w*AlbB, and *w*Au are more robust, retaining high densities when larvae are reared at cyclical temperatures of 26–37°C. These laboratory studies demonstrate the potential for heat stress to affect the success of *Wolbachia* interventions, but conditions experienced by mosquitoes in field situations are more complex than in an incubator (Ross et al., 2019).

Jiggins (2017) noticed that if *Wolbachia* infection is costly to the insect, then at low prevalence, these costs may overshadow the benefits of CI and *Wolbachia* will be lost from the population. The same takes place if infected females do not transmit *Wolbachia* to all their offspring. This results in a threshold prevalence beyond which *Wolbachia* will invade the population and below which it will be lost. Release programs must be adequately large that this threshold is surpassed, or else, the infection will be lost when the releases stop. In *A. aegypti*, *Wolbachia* has a near-perfect vertical transmission. On the other hand, the infected females are estimated to suffer a fitness cost of about 20%, mostly due to a reduction in their fertility. This results in a threshold prevalence of about 20%–30%, which must be reached by releasing infected mosquitoes, and after this point, *Wolbachia* will carry on to increase in prevalence without additional involvements, at least in isolated populations. The effect of *Wolbachia* presence on reproductive success is illustrated in Figure 11.2.

This reproductive advantage depends on the prevalence of *Wolbachia* in the population, since when *Wolbachia* is rare, females are not likely to mate with infected males. The *Wolbachia*

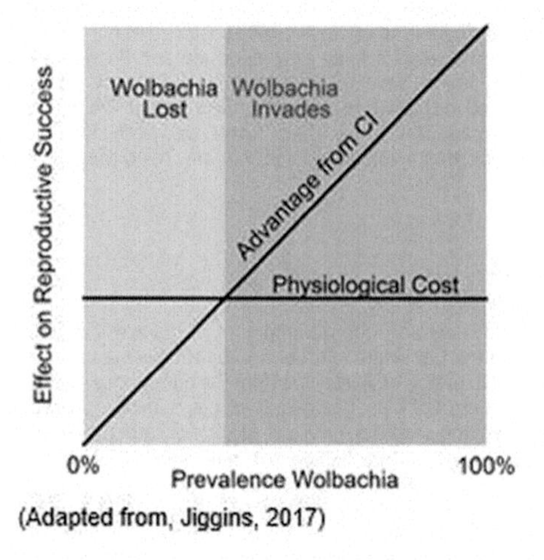

Figure 11.2 The effect of *Wolbachia* presence on reproductive success.

strain in *Aedes aegypti* transmits a physiological cost, reducing the fecundity of infected females. If this cost exceeds the value of CI, then the infection is lost from the population. This makes a threshold prevalence below which *Wolbachia* is lost and above which it invades the population.

Practical Applications of *Wolbachia*

Various projects have been deployed, including the World Mosquito Program (WMP), in various countries all over the world, such as in Australia, Vietnam, Colombia, and Brazil, to take advantage of *Wolbachia* in combating vector-borne diseases through population modification and suppression (Carrington et al., 2017; Macias et al., 2017).

Evidence suggests *Wolbachia* could be an effective means to limit transmission of the Zika virus, as the bacteria confers virus resistance in vectors. Some of such evidence has been gathered from Australia in the case of dengue and Brazil for Zika (Caragata and Moreira, 2017). Verily announced a plan to release about 20 million *Wolbachia*-infected *Aedes aegypti* mosquitoes in California to combat the Zika virus. Singapore's National Environment Agency has been working on a project to release male *Wolbachia* mosquitoes to suppress the urban *Aedes aegypti* mosquito population and fight dengue (Caputo et al., 2018; Caragata et al., 2019; Tan et al., 2017). MosquitoMate is another American project is that has embraced *Wolbachia* to release infected males that exhibit CI on mating with wild-type females (Macias et al., 2017).

Wolbachia Used to Prevent Disease

Naturally, existing strains of *Wolbachia* have been revealed to be a means for vector control strategies because of their occurrence in arthropod populations, such as mosquitoes. Owing to the distinctive traits of *Wolbachia* that cause CI, some strains are beneficial to humans as a promoter of genetic drive within an insect population. *Wolbachia*-infected females are capable of producing offspring with uninfected and infected males; yet uninfected females are only able to produce viable offspring using uninfected males. This provides infected females with a reproductive advantage that is superior to the higher frequency of *Wolbachia* in the population. Computational models calculate that introducing *Wolbachia* strains into natural populations will decrease pathogen transmission and decrease overall disease problems. An example consists of a life-shortening *Wolbachia* that can be used to control dengue virus and malaria by eradicating the older insects that have more parasites. Encouraging the survival and reproduction of younger insects reduces selection pressure for the evolution of resistance.

Wolbachia has also been acknowledged to prevent the replication of the chikungunya virus (CHIKV) in *A. aegypti*. The *w*Mel strain of *Wolbachia pipientis* significantly reduced infection and spreading rates of CHIKV in mosquitoes compared to *Wolbachia* uninfected controls and a similar phenomenon was witnessed in yellow fever virus infection, transforming this bacterium in an exceptional promise for YFV and CHIKV suppression.

Wolbachia also prevents the secretion of the West Nile virus (WNV) in cell line Aag2 derived from *A. aegypti* cells. The mechanism is to some extent novel, as the bacteria in reality improves the production of viral genomic RNA in the cell line *Wolbachia*. Also, the antiviral outcome in intrathoracically infected mosquitoes rests on the strain of *Wolbachia* and the replication of the virus in orally fed mosquitoes was entirely inhibited in the wMelPop strain of *Wolbachia*.

Wolbachia may induce reactive oxygen species-dependent activation of the Toll (gene family) pathway, which is vital for activation of antimicrobial peptides, defensins and cecropins that help inhibit virus multiplication. Equally, certain strains dampen the pathway, leading to higher replication of viruses. One example is with strain *w*AlbB in *Culex tarsalis*, where infected mosquitoes essentially carried the West Nile virus (WNV) more often. This is because *w*AlbB obstructs REL1, an activator of the antiviral Toll immune pathway. Consequently, careful studies of the *Wolbachia* strain and ecological concerns must be done before releasing artificially infected mosquitoes in the environment.

Deployments of *Wolbachia*

In 2016 it was suggested to fight the spread of the Zika virus by breeding and releasing mosquitoes that have purposely been infected with an appropriate strain of *Wolbachia*. A present-day study has shown that *Wolbachia* have the capacity to block the spread of Zika virus in mosquitoes in Brazil. In October 2016, it was broadcasted that US$18 million in funding was being assigned for the use of *Wolbachia*-infected mosquitoes to fight Zika and dengue viruses. Deployment was scheduled for early 2017 in Colombia and Brazil (Schinirring, 2016).

In Brazil, these mosquitoes have been released in two suburbs of Rio de Janeiro since late 2014, as part of the preliminary description of the appropriateness of the bacterium for use in Latin American mosquito populations. Given the latest occurrence of Zika in the region, it was essential to understand whether wMel caused a comparable inhibition of ZIKV infections in Brazilian mosquitoes. A pilot experiment of oral infections using two presently circulating ZIKV isolates (BRPE and SPH) that were isolated from patient blood during the late 2015 outbreak was done.

12

CRISPR Gene Drives

History

The CRISPR (clustered regularly interspaced short palindromic repeats)-Cas (CRISPR- associated proteins) system is a gene-editing tool that mimics an adaptive immunity in prokaryotes to defend against invasive genetic elements in form of viruses and plasmids (Davidson et al., 2020; Faure et al., 2019; Han and She, 2017). This bacterial immunity mechanism was revealed in the 1990s and early 2000s during genome sequencing efforts of archaeal and bacterial genomes. Comparative genomics on bacteria showed that the system comprises foreign origin spacers, diverse Cas proteins with conserved RNA binding and helicase and/or nuclease domains. The findings were experimentally tested and revealed the system's nature as providing an RNA-guided antiviral immunity. The immunity is gained upon exposure to invading genetic elements (Barrangou and Horvath, 2017; Ishino et al., 2018; Murugan et al., 2017).

Peculiar CRISPR repeats were first identified in 1987 when *Escherichia coli* was sequenced for a gene coding for an alkaline phosphatase isozyme. The repeat sequence contained five 29-nucleotide repeats of identical sequence that are interspaced by 32-nucleotide unique sequences, which is a part of the 12 repeat locus clustered with the CRISPR-Cas system in *E. coli*, as shown in a paper published by Ishino et al. (Han and She, 2017; Ishino et al., 2018). This study became the cornerstone on which other studies were built. Table 12.1 shows various studies whose results enhanced the understanding and use of CRISPR-Cas9 as a genome editing tool. After the discovery of repeated sequences in 1987, various names were used for the phenomenon including, multiple direct repeats (DRs), short regularly spaced repeats (SRSR), and large clusters of tandem repeat (LCTR). The name CRISPR was coined by Jansen and Mojica around 2001. In the same study, the CRISPR-associated (Cas) genes were also identified, and it was found that the four genes (Cas1–4) are located in clusters in the immediate proximity of CRISPR loci (Han and She, 2017). Makarova et al. found out that these genes code for nucleases and helicases. These proteins were named repair-associated mysterious proteins (RAMPs). By 2006, more than 50 families of Cas proteins had been discovered to be part of the CRISPR system variations (Brandt and Bucking, 2019; Barrangou and Horvath, 2017; Ishino et al., 2018; Murugan et al., 2017).

Target Cells

The CRISPR-Cas9 system is inherently present in 90% of archaea and 40% of prokaryotic cells, where it provides immunity against previously encountered invading nucleotides (Davidson et al., 2020; Faure et al., 2019). However, it also works in the eukaryotic cell, provided the machinery with endonuclease and gRNA have been inserted into the cell. The CRISPR-Cas9 system is intrinsically adapted to prokaryotic cells that lack compartmentalization. The challenge with deploying CRISPR in eukaryotic cells is that the Cas9 enzyme cannot readily enter the nucleus past the nuclear membrane (Barrangou and Horvath, 2017). Cas9 cannot edit the eukaryotic genome if it remains in the cytoplasm. Adding a nuclear localization signal (NLS) gene to the Cas9-gRNA cassette facilitates its movement into the eukaryotic nucleus of Cas9 protein to effect genome editing (Lin and Luo, 2019; Roggenkamp et al., 2018).

Ease of Use

As compared to other genome editing platforms, such as TALENs, ZFNs, and megaTALs, CRISPR-Cas9 is easily adaptable to a different application and species, as only two components need to be delivered into the cell (Carroll, 2017; Hatada, 2017). Delivery of the homing gRNA and Cas9 endonuclease can be achieved using various cell delivery methods that include both viral and

DOI: 10.1201/9781003165316-12

Table 12.1 Key Studies in the Development of the CRISPR-Cas9 System

1987	The first discovery of CRISPR XE "CRISPR" clustered repeats (Ishino *et al.*)
1989	Specific study of these clustered repeats in *Enterobacteria* (Nakata *et al.*)
1993	Use of CRISPR XE "CRISPR" for *Mycobacterium tuberculosis* typing (Groenen *et al.*)
1995	Identification of CRISPRs in *Haloferax* (Mojica *et al.*)
1996	Description of CRISPRs in *Cyanobacteria* (Mesopohl *et al.*)
1999	Use of CRISPR XE "CRISPR" for *Streptococcus pyogenes XE "Streptococcus pyogenes"* typing (Hoe *et al.*)
2000	CRISPR XE "CRISPR" is widespread in archaea and bacteria (Mojica *et al.*)
2002	Proposed CRISPR XE "CRISPR" name; identification of Cas genes (Jansen *et al.*) Discovery of CRISPR XE "CRISPR" transcripts (Tang *et al.*) DNA XE "DNA" repair XE "repair" hypothesis for Cas genes (Makarova *et al.*)
2005	The foreign origin of spacers; proposed adaptive immunity function; identification of PAM XE "PAM" (Mojica *et al.*; Pourcel *et al.*; Bolotin *et al.*)
2006	Putative RNAi-based mechanism proposed (Makarova *et al.*)
2007	First experimental evidence of CRISPR XE "CRISPR" adaptive immunity; Cas9 XE "Cas9" is required for the immunity type II system (Barrangou *et al.*)
2008	Discovery of the proto spacer adjacent motif in *S. thermophilus* (Deveau, *et al.*; Horvath *et al.*) Matured crRNA XE "crRNA" guide interference for type I system (Brouns *et al.*) Type III-A CRISPR XE "CRISPR" system targets DNA XE "DNA" (Marraffini *et al.*)
2009	Type III-B Cmr complex cleaves ssRNA (Hale *et al.*)
2010	Cas9 XE "Cas9" cleaves target DNA XE "DNA" within protospacer XE "protospacer" via DSB XE "DSB" (Garneau *et al.*)
2011	Classification of CRISPR XE "CRISPR" systems into three types (Makarova *et al.*) Discovery of tracrRNA XE "tracrRNA" (Deltcheva *et al.*) Reconstitute type II system in a distant organism (Sapranauskas *et al.*) Discovery of seed sequence (Wiedenheft *et al.*; Semenova *et al.*)
2012	In vitro characterization of DNA XE "DNA" targeting by Cas9 XE "Cas9" (Jinek *et al.*; Gasiunas *et al.*)
2013	Genome editing by Cas9 XE "Cas9" (Cong *et al.*; Mali *et al.*; Jiang *et al.*) Type III-B Cmr mediates DNA XE "DNA" interference (Deng *et al.*)
2014	Crystal structure of Cas9 XE "Cas9" (Jackson and Nishimasu *et al.*; Anders and Jinek *et al.*) Cas9 XE "Cas9"-based eradication of latent HIV XE "HIV" infection (Hu *et al.*)
2015	Type III-A Csm cleaves transcribed dsDNA in vitro (Samai *et al.*) Crystal structure of a Chimeric Cmr complex (Osawa *et al.*) Cpf1 is a single RNA XE "RNA"-guided nuclease XE "nuclease" (Zetsche *et al.*)
2016	Cmr and Csm mediate RNA XE "RNA" activated DNA XE "DNA" cleavage XE "cleavage" (Elmore *et al.*; Estrella *et al.*; Kazlauskiene *et al.*; Han *et al.*) Discovery of anti-CRISPR XE "CRISPR" proteins XE "proteins" against Cas9 XE "Cas9" in *N. meningitides* (Pawluk *et al.*)

Sources: Barrangou and Horvath, 2017; Han and She, 2017.

nonviral methods. A method that is gaining popularity is ReMOT for the delivery of RNP into insects for gene editing (Razzaq and Masood, 2018).

Efficiency

CRISPR-Cas9 gene drives are highly efficient in performing gene edits. High efficiency of about 97% has been recorded in some studies using the CRISPR gene drive construct. CRISPR gene drive efficiency can be measured at different levels including sequence targeting, endonuclease fidelity, expression of the protein, and rate of spread of a gene in a population (Champer et al., 2017; Jiang and Doudna, 2017; Khan et al., 2018). In some cases, the rate of spreading genetic modification in the population is very low. This may be explained by the reduced fitness of transgenics because of the metabolic cost related to the expression of the inserted Cas9 and gRNA (Nash et al., 2019).

The resistance to CRISPR gene drives can be attributed to the high prevalence of the NHEJ repair mechanism over HDR (Unckless et al., 2017). With HDR, the Cas9-gRNA cassette is successfully inserted in the genome, leading to successful genetic editing. However, NHEJ causes the two ends of DSB to join back together and is coupled with very high chances of indels leading to mutations at the locus. Such mutations prevent successful homing of the RNP at the locus; hence, the spread of intended genetic change is inhibited. For this reason, the CRISPR-Cas9 gene drives catalyze resistance to themselves due to their susceptibility to NHEJ-induced mutations, a phenomenon also likened to drive "braking". Populations are inherently characterized by a genetic variation that may lead to a reduction in the efficiency of CRISPR gene drives as the ribonucleoprotein complex fails to target the intended locus for editing (Champer et al., 2017; Parmentier et al., 2009; Unckless et al., 2017).

Specificity

The platform can be used in different organisms with little modification, involving the design of a specific gRNA that matches the sequence in the genome. CRISPR has been experimentally proven to work in bacteria, fungi, arthropods, and mammals, including humans. The specificity of each gene drive experiment is determined by the hybridization between the target sequence and gRNA, including the system-specific PAM sequence (Chen, 2019; Jiang and Doudna, 2017). CRISPR is considered very specific in many different organisms.

Error Rate

CRISPR platform is prone to off-target errors. This occurs in gene-editing endeavors as much as it may happen in gene drive projects. Gene drives facilitated by the CRISPR platform are prone to the NHEJ repair mechanism, with inherently high rates of mutations emanating from indels (Jiang and Doudna, 2017; Lin and Luo, 2019; Razzaq and Masood, 2018). Strategies to reduce CRISPR error rate include the use of double checkpoints for multiplexed targeting of the locus; use of truncated gRNA with less than 20 nucleotides to destabilize binding at off-target sites; use of high fidelity Cas9 mutants such as SpCas9-HF1, eSpCas9, or HypaCas9; use of bioinformatics tools to predict potential off-target activity where there is sequence homology in the genome; use of paired Cas9 nickases with one of the cutting domains inactivated; and use of a fused dCas9-FokI nuclease (Chapman et al., 2017; Chen, 2019; Fleming et al., 2018; Hajiahmadi et al., 2019).

Principle

In bacteria, the CRISPR system consists of two elements:

1. An array comprising variable spacers that are primarily derived from foreign genetic elements separated by a palindrome that is repeated throughout the array, referred to as CRISPR locus.
2. Operons of Cas genes encoding proteins that have endonuclease and helicase properties.

Transcription of CRISPR locus yields a long transcript termed pre CRISPR RNA (pre-crRNA) that is processed into small crRNAs of the single spacer-repeat unit (mature crRNAs) by RNAse III, whereas expression of the Cas gene operons produces Cas proteins. Then, crRNAs and Cas proteins form ribonucleoprotein (RNP) complexes that achieve the immunity by specifically recognizing invading genetic elements via sequence complementarity between the crRNA and its corresponding DNA sequence (protospacer) on invading genetic elements and targeting it for destruction. The protospacer region targeted by crRNA is usually flanked by a protospacer adjacent motif (PAM), usually composed of two to four nucleotides. The PAM sequence is specific for a given CRISPR-Cas system. In most cases, PAMs are crucial in allowing CRISPR-Cas systems to distinguish self from non-self and prevent the binding of CRISPR complexes to the CRISPR arrays from where they originated. Once foreign nucleic acids are bound, the CRISPR-Cas complex mediates their cleavage and subsequent destruction (Davidson et al., 2020; Faure et al., 2019).

The defense function of the CRISPR system in bacteria and archaea involves three stages:

1. **Adaptation**, whereby DNA segments (called protospacers) from foreign genetic elements (such as viruses and plasmids) are integrated into a CRISPR array/locus.
2. **CRISPR RNA (crRNA) maturation**, whereby a CRISPR array is expressed as a pre-crRNA (long RNA molecule containing multiple spacers inserted between repeats) and cleaved into mature crRNAs that contain the spacer flanked by a portion of an adjacent repeat by a non-Cas RNAse (RNAse III).
3. **Interference**, during which effector Cas nucleases complexed with the crRNA target and cleave cognate DNA molecules (Faure et al., 2019).

It has already been established that the CRISPR endonuclease Cas9 can cleave double-stranded DNA, causing breaks away from the target site and that gRNA can also be retargeted or reprogrammed, specificity of the construct is highly crucial. A variation of constructs exists as described in the following.

Autosomal Construct—Here, the gene drive construct is considered as a single allelic element, which is denoted by the capital letter, for instance, W. If the construct is not present at the corresponding locus, the denotation w is applied.

X-Linked Construct—First, this applies to organisms displaying heterogametic sex for the males, that is, XY (for example, the male dogs). In this case, the gene drive construct is placed on the X, given that the male of the selected target population is heterogametic. Predictability is not easy for such a setup, as the females carry double XX instead of XY.

Y-Linked Construct—This is a case in which the construct is inserted at a place on the Y chromosome.

The work by Jennifer Doudna and her team has since modified this system so that it is amenable to experimental conditions by fusing the tracrRNA and crRNA to create a homing gRNA. This has led to the reduction of CRISPR components to two (gRNA and Cas9) as compared to four that are inherent in bacteria.

Genome editing involves insertion, deletion, or replacement of a gene or nucleotide sequence for eliminating or inducing specific and desired characters in the genome. This can be done to achieve various ends such as mutating a gene and evaluate the effect of different alleles, deleting a gene from the genome and inserting a desired foreign gene into the genome so that it can be expressed (transgenesis) (Razzaq and Masood, 2018). In situations where transgenesis should take effect in the entire population, two strategies may be used: population replacement and population suppression (Nash et al., 2019; Unckless et al., 2017; Wedell et al., 2019). CRISPR-Cas9 gene drives are used to rapidly spread the desired transgene or genetic modification in a population. However, the introduction of a transgene presents a fitness cost to the organism. This fitness cost emanates from the metabolic processes related to the expression of the coded protein(s). For instance, the fitness cost may lead to reduced fecundity in the transgenic as compared to the wild-type organism (Oberhofer et al., 2018; Ross et al., 2019; Unckless et al., 2017). The fitness cost associated with transgenesis presents a challenge when we are faced with an effort to spread a transgene in a population. Natural selection in this case will select wild-type individuals in favor of the transgenics because their environmental fitness is compromised. When Mendelian laws are left to determine the future generation after the introduction of transgenics, there is a tendency for the population to revert to become composed of all wild types (Wedell et al., 2019). However, the low fitness in the transgenic individuals also presents an opportunity for a population suppression strategy. This may involve homing to the germline and would cause loss-function of genes leading to nonviability or sterility (Oberhofer et al., 2018).

CRISPR gene drives bias inheritance so that genetic modification has a greater than 50% chance of being passed on to offspring. They defy Mendelian inheritance laws (Champer et al., 2017; Oberhofer et al., 2018; Roggenkamp et al., 2018). Gene drives can also be looked at as selfish genetic elements that manipulate gametogenesis, embryogenesis and other reproduction processes to increase the prevalence of their transmission to future generations. They have been commonly used to control the population of invasive organisms and control the population of disease vectors and pests through suppression or replacement. CRISPR uses HEGs as a vehicle for creating and, at the same time, spreading transgenes in populations (Oberhofer et al., 2018; Wedell et al., 2019).

Generally, the CRISPR gene drives require the following components:

1. Cas9 endonuclease for cutting DNA. There are several different Cas enzymes, but the most common is Cas9 from *Streptococcus pyogenes*
2. gRNA for targeting and homing to a specific sequence (locus) in the genome
3. DNA sequence template for insertions during HDR

Cas9 endonuclease cuts both of the complementary DNA strands using RuvC and HNH domains to make a DSB. When Cas9 performs endonuclease activity at a targeted location in the genome, the cell machinery may repair the DSB in two ways:

1. Nonhomologous end joining (NHEJ)
2. Homology-directed repair (HDR)

The NHEJ repair mechanism may lead to indels of nucleotides at the blunt ends of the DSB. Indels are a source of unintended mutations at the location targeted by gRNA in the genome, resulting in low efficiency of the system. On the other hand, HDR uses a template with complementarity to the flanking regions of location targeted for genome editing. The template directs the sequence of nucleotides to be added to the DSB location. HDR is used to insert a desired gene sequence in the genome (Lin and Luo, 2019; Oberhofer et al., 2018). The transgenes or orthologous sequences added can be from other species (Frieß et al., 2019).

A nonhomologous repair of the cut site by NHEJ or microhomology-mediated end joining (MMEJ) decreases the conversion rate, as these alternative mechanisms often cause mutations or deletions at the target site of the endonuclease. Depending on the genomic location, HDR vs. NHEJ efficiency might be as low as ~10% (Frieß et al., 2019). To lessen these actions, CRISPR-Cas9 can be used to increase HDR gene expression and suppress NHEJ genes. This can be accomplished by the inclusion of HDR genes and NHEJ-repressor genes. Also, the generation of nucleases generating sticky-end overhangs as opposed to blunt ends may improve the repair in the target organism. The rate of HDR depends on the species, cell type, developmental stage and cell cycle phase. For instance, exact copying was attained with up to 97% efficiency in mosquitoes but only 2% in fruit flies (Frieß et al., 2019).

A gene drive organism transmits the gene drive cassette and mates with a wild type. The gene drive cassette expresses the CRISPR-Cas9 complex, which then cuts its recognition site defined by the gRNA on the homologous chromosome. This cut can then either be repaired by HDR copying the gene drive cassette into the cut region or by mechanisms producing a homing resistant allele, like NHEJ.

Initial observations propose that resistance can also be the result of a "dominant maternal effect" (Ping, 2017). It is anticipated that Cas9 deposits in the oocyte cause early cuts in the genome of the sperm cell, soon after fusion. Upon fertilization, if enough Cas9 (and gRNA) is in the cytoplasm of the zygote that is homo- or heterozygous for the GD the CRISPR-Cas complex finds and cuts its recognition sites in the sperm's genome before the homologous female genome is sufficient to be employed for homologous recombination. Devoid of a homologous template, the cuts are then repaired by NHEJ and hence a resistant allele may ascend. In such an occasion, the amount of gRNA variants is worthless. The proliferation of resistant individuals may be decreased by targeting essential genes (Noble et al., 2017).

Procedure

The principle of gene drives has often been applied to the cause of suppressing the population of disease vectors and pests, including building parasite resistance in vector populations. The organisms that have been used for such studies include mosquitoes (*Anopheles* spp., *Culex* spp., and *Aedes* spp.) and fruit flies (*Drosophila* spp.) among others (Frieß et al., 2019). It is then imperative that a lot of literature has been created related to the procedure for gene drives in insects. To modify the genome of these insects, a ribonucleoprotein complex is needed, together with a template in the case of gene insertion using HDR (Kyrou et al., 2018). CRISPR can also be engineered to cause insertions or deletions (Doudna and Charpentier, 2014). This procedure is more effective when performed during vitellogenesis and early stages of embryo development rather than later stages (Heu et al., 2020).

In a heterozygous animal expressesing the Cas9 nuclease, the gRNA targets cleavage of the wild-type homologous chromosome. Genomic sequences that flank the active genetic element will then correct the DSB by HDR, which copies the active genetic element from the donor to the receiver chromosome and converts the heterozygous genotype to homozygosity. The frequency of spreading the active genetic element to the next generation is for that reason greater than expected by random segregation of heterozygous alleles and is called "super-Mendelian" (Frieß et al., 2019).

The common method of delivery of HEG into the developing embryo is microinjection, which should target the germplasm found in the posterior part of the egg. The germplasm develops into part of the insect anatomy responsible for the production of gametes; hence, any genetic modification can be transmitted to the next generation. Progeny testing is done to ascertain the success of genome editing (Chaverra-Rodriguez et al., 2018; Xu et al., 2020).

To deliver RNP into the germplasm, two plasmids are used:

1. A plasmid that has a Cas9 gene and germline promoter such as nanos so it can be expressed during the early stages of vitellogenesis.
2. A plasmid that has a gRNA gene that can be transcribed to produce the homing transcript and also germline promoter.

This method is delicate, and the success rate of microinjection is very low. It is also slow, as the process follows many processes, including transcription, transcript processing, translation and post-translation processing of protein. However, it is easy to use, as it relies on cell machinery (Heu et al., 2020; Xu et al., 2020).

Receptor-Mediated Ovary Transduction of Cargo (ReMOT Control) is an alternative method. The yolk uptake signal is combined with the Cas9 gene and delivered into the hemolymph. The receptors in the ovaries will then activate the loading of the complex into the developing eggs. The procedure is simple, as it involves injecting RNP complexed with yolk uptake signal into any part of the insect body into the hemolymph. High efficiency of gene editing has been reported for the maternal chromosomes as compared to that from paternal chromosomes when fertilization uses sperm from the spermathecae (Chaverra-Rodriguez et al., 2018; Heu et al., 2020; Macias et al., 2020).

Mutagenic Chain Reaction (MCR)

One way to rapidly spread a recessive transgene in a population is through the use of a mutagenic chain reaction (MCR). Since in such a scenario the desired phenotype cannot only manifest in a homozygous state, the system must generate autocatalytic mutations to produce homozygous mutations. MCR mutations efficiently spread from their chromosome of origin to the homologous chromosome, thereby converting heterozygous mutations to homozygosity in the vast majority of somatic and germline cells (del Amo et al., 2019; Schaap, 2018; Unckless et al., 2017).

Mutagenic Chain Reaction Construct

MCR insertional mutants could be generated with a construct having three components: Cas9 endonuclease, gRNA and homology arms (HAs) flanking the Cas9-gRNA cassettes. These HAs should match the two genomic sequences adjacent to either side of the targeted loci. The gene of interest is cloned into a plasmid vector then modified as desired using site-directed mutagenesis or other nucleic acid manipulation methods. The construct is then inserted into the genome of an organism

The expression of the Cas9-gRNA cassette inserted in the genome results in the production of Cas9 nuclease and gRNA, which combine to form ribonucleoprotein (RNP). The RNP targets the gene of interest and cuts the DNA. The cell machinery may use the HDR mechanism to repair the DSB on the genomic locus. Since most organisms are diploid or polyploid, the same mechanism is used to modify the homologous chromosomes at the corresponding loci as long as the targeted sequence exists. MCR gene drives should rapidly spread the transgene or genetic modification in the population and show features of super-Mendelian inheritance (Oberhofer et al., 2018).

Versions

There are two kinds of gene drives that commonly use the CRISPR platform. The first one is the modification drive, which is used to alter wild-type alleles. Modification of the genome is ideally done in the early zygote stage so that it can perpetuate to future generations. Modification gene drive may also lead to loss-of-function, thereby affecting the behavior of an organism. The second kind is the suppression gene drive, which is used to reduce the population density or eradicate the organism. The gene drives can be deleterious to the organism and result in death, reduce fecundity, or cause sterility (Borsenberger et al., 2018; del Amo et al., 2019; Nestor and Wilson, 2020).

Advantages

Generally, CRISPR-Cas systems provide a uniquely powerful gene-editing tool that is readily adaptable to newly encountered genomes. Hence, there is no need for the prior species-specific study. This has resulted in the widespread use of CRISPR (Davidson et al., 2020). CRISPR is characterized by a high on-target mutation rate (targeting efficiency), flexibility, less cost, simplicity, and high-efficiency multiplexing loci editing and is highly versatile for a variety of applications (Chen, 2019; Hajiahmadi et al., 2019; Oberhofer et al., 2018).

Disadvantages

CRISPR gene editing is known to cause off-target effects that produce unintended negative effects. These off-target effects may arise from the gene-editing machinery targeting a similarly homologous sequence at a different location in the genome. The effects may lead to terrible genetic diseases causing abnormality or even death (Borsenberger et al., 2018; Jiang and Doudna, 2017; Lin and Luo, 2019). There is also substantial risk associated with this highly invasive method. If there is a spontaneous change of the gene drive cassette, it may be difficult to curb a rapid spread of the unintended mutation in the population whose effects may be disastrous to ecosystems.

There are also potential risks of synthetic or mutated gene drivers invading nontarget species or populations (Wedell et al., 2019). The HEG drives rely on the presence of a predetermined gene sequence and its efficacy is sensitive to genomic sequence variation. Species populations are usually characterized by sequence polymorphisms. Such polymorphism may also arise from mutation and as a result of DSB repair through NHEJ, which is error-prone. Sequence variants are a source of resistance to homing and hence block the spread of HEG for purposes of population suppression (Oberhofer et al., 2018).

Causes of Failure of CRISPR Gene Drives

Failure of the CRISPR-Cas9 gene drive technology could be a result of the following:

1. Reduced fitness of laboratory-reared GM insects because of inbreeding (Frieß et al., 2019).
2. DNA repair via NHEJ and MMEJ instead of HDR.
3. On-target mis-insertions; sometimes gRNA sequences are inserted (Li et al., 2020).

4. Inadequate or defective copying during HDR (if the deletion conserves the reading frame, it leads to a homing-resistant allele) (Marshall et al., 2017).
5. Off-target effects (undefined binding of gRNA causes accidental insertions at different locations).
6. On-target mis-insertions (undesirable genes or gene fragments are inserted into the target locus instead of or additional to the anticipated genes); occasionally, gRNA sequences are inserted.
7. The occurrence of homing-resistant alleles owing to random target site mutagenesis.
8. Sequence polymorphisms (resistance due to genetic variations within a species). To overcome this problem, multiple gRNA alternatives can be added to the CRISPR-Cas cassette.
9. Intragenomic interactions (the distance of gRNA target sites may affect homing rates (Marshall et al., 2017).
10. Release of phenotypic wild types carrying the nonfunctioning construct (would decrease the suppressive outcome and might constitute a persistently GD-resistant subpopulation).

As a prospective method to reverse harmful damages caused by CRISPR-Cas9 gene drives, the (mass) release of a secondary drive can be used. It can be a rescue drive that cuts out the cargo gene and forms a resistant locus (Frieß et al., 2019). This, on the other hand, would also mean a population replacement, persistently introducing extra synthetic genetic material into the ecosphere, as the CRISPR-Cas9 system would persist in the populations' gene pool. Some suggested drives to limit the spread are:

1. Reversal (or overwriting) drive
2. Immunizing drive (preemptively) to make populations immune to another drive
3. Split drive to serve local confinement
4. Daisy chain drive for confinement in space and time

Applications of CRISPR Gene Drives

CRISPR Cas9 gene drives are used for the eradication of vector-borne diseases in humans, crops, and livestock. Common diseases that have attracted scientists to research the use of CRISPR drives include malaria, dengue, chikungunya, Zika, and many others. Significant research efforts have been channeled toward the eradication of malaria, a disease caused by a plasmodium parasite that is transmitted by *Anopheles* mosquitoes. Strategies for malaria eradication include a modification gene drive that spreads a malaria-resistance gene and suppression gene drive that causes sterility in female *Anopheles* that are homozygous for a recessive gene (Emerson et al., 2017; Hammond and Galizi, 2018).

Gene drives are also used for population control of invasive species and crop and livestock pests. This is achieved mostly by population suppression gene drives. Such efforts seek to minimize the impact of these organisms on the ecosystems where they exist. Examples of studies that sought to eradicate invasive vertebrates include organisms such as mice, rats and rabbits, especially on islands. Loss-of-function gene drives can also be implemented as a modification strategy to reduce the ecological impact of invasive organisms (Emerson et al., 2017; Noble et al., 2017; Prowse et al., 2017). Gene drives have also been used to reduce the population size of agricultural pests. The CRISPR-based drive has been shown to reverse resistance of insect and weed pests to pesticides by resensitizing pests to toxins (Neve, 2018).

Special gene drives are also being designed to help protect endangered species through modification genomes so that they can resist diseases with the potential to wipe out the entire population. An example is drives to protect frogs and other amphibians that are in dramatic decline worldwide due to chytrid fungus. The fungus causes a skin disease that is often lethal. A gene preventing fungal infections could potentially save many frogs and other amphibian species from extinction (Norwegian Biotechnology Advisory Board, 2017; Rode et al., 2019; Roggenkamp et al., 2018).

Target Malaria is a nongovernmental organization working on projects to reduce the spread of malaria through CRISPR gene drives that suppress mosquito populations and transgenes for *Plasmodium* parasite resistance. They use self-sustaining gene drives for either malaria resistance, inability to find a blood source, or female infertility. Another technique they use involves removing a chromosome that determines sex, leading to all offspring being male (Target-Malaria, 2020).

The Oxitec mosquito project in Brazil aims at suppressing the mosquito population using a "lethal" gene. The self-limiting gene is inserted into male *Aedes aegypti* mosquitoes using CRISPR. When the male mosquitoes are released, they mate with females to produce offspring that are all heterozygous. As heterozygotes, the gene is expressed and self-regulates, since the tTAV protein sets up a positive feedback loop by triggering overproduction of itself. The over-expression of this gene leads to the overproduction of a protein at the expense of other essential cellular processes. As a result, the heterozygous mosquitoes cannot reach the adult stage that gives them a chance to reproduce. Tetracycline is added to the diet of the released male mosquitoes to block the regulatory activity of tTAV protein; hence, they can survive and reproduce (Jones et al., 2019; Romeis et al., 2020).

Examples of How CRISPR Gene Drives Can Be Used

Construct

Champer et al. (2017) designed two CRISPR-Cas9-based gene drive constructs targeting the X-linked gene in *D. melanogaster* to reveal insights into mechanisms of resistance allele formation. Disruption of this gene causes a recessive phenotype, specified by a lack of dark pigment in adult flies.

The first drive construct contains a Cas9 endonuclease gene driven by the nanos promoter, with a gRNA targeting the coding sequence of the gene. In this case, they expect that most resistance alleles caused by a mutated target site should disrupt the gene (r2 resistance alleles), whereas resistance alleles that preserve the function of r1 resistance alleles should occur less frequently.

Their second drive construct contains a gRNA targeting the promoter of the yellow gene and a copy of Cas9 driven by the vasa promoter. In this situation, resistance alleles have to be primarily of type r1. The precise insertion site of the promoter was selected to induce a yellow phenotype in the wings and body when interrupted by a large construct but to maintain male mating success, which is reduced when the coding site or downstream regions of the promoter are interrupted by a large construct.

Both constructs also encode a dsRed protein, driven by a 3xP3 promoter, which produces a simple recognizable fluorescent eye phenotype (R) that is dominant and permits us to identify the existence of drive alleles in individuals.

13

The Killer-Rescue System

History

Killer-rescue (K-R) systems were first identified in bacteria (Ogura and Hiraga, 1983). The K-R system was first proposed by Gould et al. (2008), and it consists of two unlinked loci, one encoding a toxin (killer allele) and the other an antidote (rescue allele). This is a system of killing and rescuing that is used by the K-R system to act antagonistically and bias their co-inheritance. The toxin and antidote could consist of miRNAs and a recoded gene or a toxic protein and toxin-inhibiting enzyme. A cargo gene can be fused to the antidote gene. K-R is a gene drive system. Self-limiting gene drive systems that have been developed and proposed include the daisy chain gene drive, one- or two-locus underdominance and toxin antidote systems including *Medea*, CleaveR and killer-rescue systems.

The K-R self-limiting gene drive, also known as K-R gene drive, is a gene drive system that does not drive toward non-Mendelian inheritance like Cas9-based homing drives. However, K-R drives through the mortality of individuals with specific alleles and this results in population replacement. A killer-rescue system gene drive involves the genetically engineered insects containing killer genes on one chromosome and rescue genes with a linked anti-pathogen gene on another separately segregating chromosome. These insects will be double homozygous for the rescue and killer, meaning that all the offspring of the matings with the wild-type population will be double heterozygous for killer and rescue. Future generations that will contain different combinations of killer and rescue will survive, but any insects that inherit only the killer will not survive to adulthood. Therefore, the only insects contributing to the next generation are those that have inherited the rescue with the linked anti-pathogen.

Mechanism

Homozygous carriers for both genes are mass-released into wild populations. Offspring inheriting the killer allele but not the rescue allele will be nonviable. Since both alleles are not linked in their inheritance, the killer allele will be quickly selected from the population, while the rescue allele confers a clear fitness gain and will increase in its prevalence. As soon as the killer allele completely disappears from the population, the rescue allele's fitness gain will also disappear. As a consequence, the rescue allele will again decline in its prevalence, unless the cargo gene confers a gain in fitness. The K-R system is highly dependent on the fitness of the rescue and cargo genes, which determines the time until the cargo genes are eliminated from a population (Marshall et al., 2017).

The K-R system utilizes the well-characterized Gal4/UAS binary expression system and the Gal4 inhibitor Gal80. Death is due to overexpression of Gal4, which is rescued by Gal4 activation of the Gal80 inhibitor in flies that have both UAS-Gal4 (K) and UAS-Gal80 (R) transgenes.

Mechanism

A killer-rescue construct is placed into the pest genome containing a maternally deposited lethal toxin, a zygotically expressed antidote and an effector (the cargo). Female pests deposit a lethal toxin into their entire progeny. However, those that inherit the *Medea* construct are protected by a neutralizing antidote expressed in the early embryo. Wild-type progeny of *Medea* females are killed and therefore the viable offspring of *Medea* females will always contain the *Medea* element. The offspring of *Medea* males mated to wild-type females are unaffected because the toxin is not expressed in males. If the killer-rescue constructs are introduced at a sufficiently high frequency in a population, they will replace the wild-type genotype and bring with them the effector.

DOI: 10.1201/9781003165316-13

When a female *Medea* heterozygote mates with a wild-type male, half of the offspring will die and the other half inherit the element, whereas when she is homozygous, all of her offspring survive and inherit the element. Males can transmit the *Medea* element without any progeny selection. As such, the *Medea* element will increase its frequency relative to wild-type and drive through a population if it is seeded above a critical threshold. The reciprocal version, "inverse *Medea*", relies upon maternal deposition of the antidote followed by zygotic killing (Marshall and Hay, 2011). This latter strategy is far less able to drive through a population and acts to sterilize the heterozygous offspring of wild-type mothers. It has been proposed, along with synthetic underdominance (Gould et al., 2008), as a contained gene drive to be used for geographically limited suppression of wild vector populations, requiring a high release threshold to spread. The *Medea* killer-rescue system is not suitable for suppression of a population because the load exerted on it will quickly dissipate as the drive spreads to fixation. Instead, these systems have been proposed as a means to drive anti-pathogen effectors through a population.

To make an effective killer-rescue system, the rescue must be expressed in at least all of the same tissues as the killer and at the same stages of development. Further, the expression of the rescue gene needs to be at sufficiently high levels to negate the lethal effects of the killer. Other considerations influence the choice of gene promoters used to drive the killer and rescue genes.

Efficiency

These systems have the potential to reduce the impact of human disease vectors or agricultural pests at the individual, community and global scale. However, systems pose challenges such as introgression into nontarget populations or movement across political borders.

Reliability of the Technology

The reliability of the technology focuses on the probability of failure of the technology with regard to its intended use. Important reliability issues are linkage-loss of the cargo gene and its driver system. The generation of resistances in the target population, coevolution of a pathogen and system decay (Alphey, 2014).

Development

In 2007, Chen et al. demonstrated the first proof-of-principle *Medea* element in which miRNA-mediated silencing of an essential maternal deposited gene was toxic to embryos that did not express the rescue transgene. Since then, two new elements have been generated along similar lines, targeting maternally deposited embryo signaling pathways and rescue by embryonic expression of a transgene source of the targeted gene (Akbari et al., 2014). As of yet, no reports have demonstrated K-R in the mosquito; however, a novel system has been proposed which may be simpler to execute—maternal deposition of CRISPR RNase targeting an embryo essential gene, can be rescued by zygotic expression of a sequence variant, obviating the requirement for an understanding of mosquito signaling pathways (Champer et al., 2016). This has the added advantage that reversal constructs can be used to halt the gene drive.

Application

The novel K-R gene drive system has been tested in *Drosophila melanogaster*. In this system, traits or genes that will either modify the pest population to get rid of its pestilent behaviors or suppress the population by disrupting a gene and reducing the average fitness of the population (such as reducing the lifespan) were assayed.

The agricultural pest *D. suzukii* is a well-known invasive fruit fly that has a negative economic impact on soft-bodied fruit production worldwide. The closely related species, *D. melanogaster*, was used to test the K-R system before transferring it into *D. suzukii* (Chiu et al., 2013). A K-R system in *D. suzukii* could be used to drive a cargo gene such as one that carries susceptibility to a certain chemical, increases susceptibility to parasitism, or decreases the overall fitness of the population and slowly causes population suppression.

Pros and Cons

The advantages of the K-R gene drive include:

1. Communities will likely feel more comfortable with initial field tests and applications.
2. It is simpler to build.
3. It is considered to have a lower risk than other gene drives.
4. Gene drive systems have the potential to be permanent and long-lasting, compared to population suppression methods.
5. These gene drive systems are economically feasible,
6. The gene drive system is less time consuming and should be considered as a long-term solution in an integrated pest management (IPM) program.

Disadvantages of this K-R gene drive system compared to a Cas9-based homing drive include:

1. It could take longer for the desirable genes to spread on a relevant timescale.
2. The cost is higher as a higher engineered-to-wild-type ratio is required for the drive at an acceptable timescale.
3. Lowered fitness of laboratory-reared GM insects due to inbreeding (colony effect).
4. The selection against the fitness burden (resistance formation or toxin inactivation).
5. Linkage loss between the rescue and cargo gene.
6. The natural evolution of an antidote or inactivation of the killer allele.

Reversibility of the K-R Gene Drive System

Gene drive systems are designed and expected to permanently spread through the target species population. These systems are often the most complex to build and deploy (Sinkins and Gould, 2006). The K-R gene drive system can be self-limiting over time and space because the system drives through the suppression of the organisms that do not inherit the rescue and linked effector. This type of gene drive is also known as a self-limiting gene drive system. Both killer and rescue genes will be lost over time, either if there are any fitness costs associated with the rescue or the release ratio is low (Gould et al., 2008). An example is where the overexpression of tetO-tTA in *Diptera* is lethal.

Comparison of K-R Systems

A comparison of the K-R systems is shown in Table 13.1.

We encourage more research into genome editing and gene drive technology to ensure the safety and protection of our resources and environment. More advocacy and awareness activities should be conducted globally.

Table 13.1 A Comparison of K-R Systems

Confinement

Gene drive	Function	Trait		Rate of spread	Release threshold	Paternal resistance	Population	Species	Region	Rate of transfer	Development
Homing	Suppression	Female infertility		5	1	4	4	5	1	4	Experimental testing
Y-drive	Suppression	Male bias		5	1	1	1	5	1	2	Early development
Homing	Replacement	Pathogen effector		5	1	4	4	5	1	4	Experimental testing
Wolbachia (CI) Killer rescue	Replacement	Survival/ immunity		3	2	3	2	5	3	2	Field testing
Medea	Replacement	Pathogen effector		3	5	3	3	5	3	3	Early development
Underdominance	Replacement	Pathogen effector		3	3	5	5	5	5	3	Experimental testing

Key:
Very low 1
Low 2
Moderate 3
High/fast 4
Very high/very fast 5

Glossary

Adeno-associated viral vectors Adeno-associated viruses, from the parvovirus family, are small viruses with a genome of single-stranded DNA. These viruses can insert genetic material at a specific site on chromosome 19 with near 100% certainty. Adeno-associated virus (AAV) vectors are the leading platform for gene delivery for the treatment of a variety of human diseases.

Allele An allele is a variant form of a gene. Some genes have a variety of different forms, which are located at the same position, or genetic locus, on a chromosome. Humans are called diploid organisms because they have two alleles at each genetic locus, with one allele inherited from each parent.

Allergenic An allergen is a type of antigen that produces an abnormally vigorous immune response in which the immune system fights off a perceived threat that would otherwise be harmless to the body. Such reactions are called allergies. Allergenic thus means capable of causing allergic reactions.

Array Gene arrays are solid supports upon which a collection of gene-specific nucleic acids have been placed at defined locations, either by spotting or direct synthesis. In literature, the term "target" can refer to either the nucleic acids attached to the array or the labeled nucleic acid of the sample.

Autosomal construct An autosome is any chromosome other than a sex chromosome. Sex chromosomes specify an organism's genetic sex. Humans can have two different sex chromosomes, one called X and the other Y. Normal females possess two X chromosomes and normal males one X and one Y. A DNA construct is an artificially constructed segment of nucleic acid that is going to be "transplanted" into a target tissue or cell. It often contains a DNA insert, which contains the gene sequence encoding a protein of interest. An autosomal construct is therefore an artificially made segment of chromosomes that are not the sex chromosomes.

B chromosomes Small extra chromosomes to the standard complement that occur in many organisms.

Bidirectional Bidirectional replication. A type of DNA replication where replication is moving along in both directions from the starting point. This creates two replication forks, moving in opposite directions.

Biotechnology Biotechnology is a broad area of biology, involving the use of living systems and organisms to develop or make products for the exploitation of biological processes for industrial and other purposes, especially the genetic manipulation of microorganisms for the production of antibiotics or hormones. It harnesses cellular and biomolecular processes to develop technologies and products that help improve lives and health.

Bioterrorism Bioterrorism is terrorism involving the intentional release or dissemination of biological agents. These agents are bacteria, viruses, insects, fungi or toxins and may be in a naturally occurring or a human-modified form, in much the same way as in biological warfare.

Bistable switch A phenomenon in which a certain threshold frequency for a gene drive system defines its eventual fate in a population. If the frequency of individuals with the gene drive in a population is above that threshold, it will spread and eventually reach fixation. If it is below that threshold, it will be eliminated from the population.

Chemokine receptors Chemokine receptors are cytokine receptors found on the surface of certain cells that interact with a type of cytokine called a chemokine. This causes cell responses, including the onset of a process known as chemotaxis that traffics the cell to a desired location within the organism.

Chimeric antigen receptor Chimeric antigen receptor are T cells that have been genetically engineered to produce an artificial T-cell receptor for use in immunotherapy. Chimeric antigen receptors are receptor proteins that have been engineered to give T cells the new ability to target a specific protein.

Cpf1 A programmable RNA-guided endonuclease from bacteria that cleaves DNA, generating staggered double-stranded breaks.

CRISPR Clustered regularly interspaced short palindromic repeats (CRISPR) is a family of DNA sequences found in the genomes of prokaryotic organisms such as bacteria and archaea. These sequences are derived from DNA fragments of bacteriophages that had previously infected the prokaryote. They are used to detect and destroy DNA from similar bacteriophages during subsequent infections.

CRISPR—Cas9 (Clustered regularly interspaced short palindromic repeats—CRISPR-associated 9). A gene-editing technology originating in bacteria that consists of an endonuclease (Cas9) and a guide RNA that can target and modify user-defined DNA and RNA sequences with great accuracy.

Cytoplasmic incompatibility Cytoplasmic incompatibility (CI) is a phenomenon that results in sperm and eggs being unable to form viable offspring. The effect arises from changes in the gamete cells caused by intracellular parasites such as *Wolbachia*, which infect a wide range of insect species.

Daisy drive gene drive A daisy drive system consists of a linear series of genetic elements arranged such that each element drives the next in the chain. The final element in the chain, which carries the "cargo," is driven to higher and higher frequencies in the population by the earlier elements in the chain. No element can drive itself. The bottom element is lost from the population over time, causing the next element to cease driving and be lost in turn. This process continues along the chain until, eventually, the population returns to its wild-type state.

Darwinian Relating to Darwinism. This according to some of Charles Darwin's theories, generally means relating to or being a competitive environment or situation in which only the fittest persons or organizations prosper.

DHFr Dihydrofolate reductase, or DHFr, is an enzyme that reduces dihydrofolic acid to tetrahydrofolic acid, using NADPH as electron donor, which can be converted to the kinds of tetrahydrofolate cofactors used in 1-carbon transfer chemistry. In humans, the DHFr enzyme is encoded by the DHFr gene.

Dimerization The chemical reaction that joins two molecular subunits, resulting in the formation of a single dimer. The process or act of forming a dimer. A protein dimer is a macromolecular complex formed by two protein monomers, or single proteins, which are usually non-covalently bound. Many macromolecules, such as proteins or nucleic acids, form dimers. The word dimer has roots meaning two parts.

DNA Deoxyribonucleic acid, one of two types of molecules that encode genetic information. DNA is a double-stranded molecule held together by weak hydrogen bonds between base pairs of nucleotides. The molecule forms a double helix in which two strands of DNA spiral about one another DNA is a self-replicating material that is present in nearly all living organisms as the main constituent of chromosomes. It is the carrier of genetic information.

DSB A double-strand DNA break (DSB) occurs or arises when both strands of the DNA duplex are severed, often as the result of ionizing radiation or nuclease activity.

Electroporation The action or process of introducing DNA or chromosomes into bacteria or other cells using a pulse of electricity to open the pores in the cell membranes briefly.

Embryonic lethality Embryonic lethality is when the embryo dies during its development in the womb. It is not born alive.

Endogenous Growing or originating from within an organism.

Endonucleases These are enzymes that cleave the phosphodiester bond within a polynucleotide chain. Some of them have no regard to sequence when cutting DNA, but many others do so only at specific nucleotide sequences. The latter group is often called restriction endonucleases or restriction enzymes.

Epigenetic It refers to external modifications to DNA that turn genes "on" or "off". These modifications do not change the DNA sequence; instead, they affect how cells "read" genes. Epigenetic changes alter the physical structure of DNA. Epigenetics involves genetic control by factors other than an individual's DNA sequence.

Eukaryotic Eukaryotic means relating to eukaryotes. A eukaryotic cell contains membrane-bound organelles such as a nucleus, mitochondria and an endoplasmic reticulum. Eukaryotes are organisms whose bodies are made up of eukaryotic cells, such as protists, fungi, plants and animals.

Feminization Feminization is the development in an organism of physical characteristics that are usually unique to the female of the species.

FokI endonuclease The restriction endonuclease FokI, naturally found in *Flavobacterium okeanokoites*, is a bacterial type IIS restriction endonuclease consisting of an N-terminal DNA-binding domain and a nonspecific DNA cleavage domain at the C-terminal.

GAL4 This is a yeast protein that regulates genes induced by galactose. GAL4 binds to 17 base pair sites referred to as the upstream activating sequences (UAS) in order to activate the GAL10 and GAL1 target genes.

Gene drive This is a natural process and technology of genetic engineering that propagates a particular suite of genes throughout a population by altering the probability that a specific allele will be transmitted to offspring (instead of the Mendelian 50% probability). The term was initially coined to describe the process of stimulating the biased inheritance of particular genes to alter entire populations. However, the term is now used to describe the actual synthetic genetic element designed to increase in frequency over time in a population. The term can be used interchangeably with the term "selfish genetic element".

Gene loci This is a specific, fixed position on a chromosome where a particular gene or genetic marker is located.

Gene promoter A gene promoter is a sequence of DNA to which proteins bind that initiate transcription of the DNA downstream of it. The gene promoter enhances expression of the sequences to which it is a promoter.

Genome editing Also known as genome engineering, or gene editing, this is a type of genetic engineering in which DNA is inserted, deleted, modified or replaced in the genome of a living organism.

gRNAs Guide RNAs (also known as gRNA, sgRNA) are the RNAs that guide the insertion or deletion of foreign DNA during genome editing.

Guided engineered nucleases Engineered nucleases are used to induce a double-stranded DNA break (DSB) at a specified locus of the gene of interest (GOI). DSBs are repaired by either nonhomologous end-joining (NHEJ) or homologous recombination (HR).

Heterodimerization This is the formation of a heterodimer, which is a protein composed of two polypeptide chains differing in composition in the order, number or kind of their amino acid residues.

Heterogametic A species containing different sex chromosomes between males and females. Humans are an example of a heterogametic species; Y chromosomes are found only in males.

Homing The process by which an endonuclease cleaves a specific DNA target sequence and copies itself, or "homes," into this target sequence. Homing utilizes the cell's homology-directed repair (HDR) machinery.

Homing efficiency The rate at which a homing-based drive gene becomes successfully copied onto the opposite chromosome via homology-directed repair.

Homing endonuclease genes Naturally occurring types of gene drives that are composed of an endonuclease encoded as a freestanding gene within introns, as a fusion with host proteins or as a self-splicing intein with the ability to home into the opposite chromosome, resulting in more than half of offspring inheriting the HEG.

Homologous Similar in position, structure and evolutionary origin but not necessarily in function; for chromosomes it implies pairing at meiosis and having the same structural features and pattern of genes.

Homology-directed repair Homology-directed repair is a mechanism in cells to repair double-strand DNA lesions. The most common form of HDR is homologous recombination. The HDR mechanism can only be used by the cell when there is a homologous piece of DNA present in the nucleus, mostly in G2 and S phase of the cell cycle.

Immunodeficiency Also known as immunocompromisation, is a state in which the immune system's ability to fight infectious diseases and cancer is compromised or entirely absent. Most cases are acquired ("secondary") due to extrinsic factors that affect the patient's immune system.

I-SceI This is a rare cutting mitochondrial DNA endonuclease involved in intron homing. It introduces a specific double-strand break in the DNA of the 21S rRNA gene and thus mediates the insertion of an intron, containing its own coding sequence into an intronless gene. It specifically recognizes and cleaves the sequence 5'-TAGGGATAACAGGGTAAT-3'.

Killer-rescue system This mechanism involves one gene that codes for toxicity (killer) and a second that confers immunity to the toxic effects (rescue).

Knock-in In molecular cloning and biology, a knock-in (or gene knock-in) refers to a genetic engineering method that involves the one-for-one substitution of DNA sequence information in a genetic locus or the insertion of sequence information not found within the locus.

Knockout A gene knockout, abbreviation: KO, is a genetic technique in which one of an organism's genes is made inoperative, "knocked out" of the organism. Knockout organisms, or simply knockouts, are used to study gene function, usually by investigating the effect of gene loss.

Lab-grown Propagated in the laboratory.

Linkers Also known as spacers, linkers are short amino acid sequences created in nature to separate multiple domains in a single protein. Often, independent proteins may not exist as stable or structured proteins until they interact with their binding partner, following which they gain stability and the essential structural elements.

Malnourishment Lack of proper nutrition, caused by not having enough to eat, not eating enough of the right things or being unable to use the food that one does eat.

Maternal-effect dominant embryonic arrest Maternal-effect dominant embryonic arrest (*Medea*) is a selfish gene composed of a toxin and an antidote. A mother carrying *Medea* will express the toxin in her germline, killing her progeny. If the children also carry *Medea*, they produce copies of the antidote, saving their lives.

Medusa A novel, threshold-dependent gene drive system capable of inducing a local and reversible population crash. Medusa consists of four components—two on the X chromosome and two on the Y chromosome. A maternally expressed X-linked toxin and a zygotically expressed Y-linked antidote result in suppression of the female population and selection for the presence of the transgene-bearing Y because only male offspring of Medusa-bearing females are protected from the effects of the toxin. At the same time, the combination of a zygotically expressed Y-linked toxin and a zygotically expressed X-linked antidote selects for the transgene-bearing X in the presence of the transgene-bearing Y. Together, these chromosomes create a balanced lethal system that spreads while selecting against females when present above a certain threshold frequency.

Meganucleases Meganucleases are "molecular DNA scissors" that can be used to replace, eliminate or modify sequences in a highly targeted way. Meganucleases are DNA cutting enzymes. By modifying their recognition sequence through protein engineering, the targeted sequence can be changed.

Mendelian Mendelian means relating to Mendel or the theories of Gregory Mendel. Mendelian inheritance refers to patterns of inheritance that are characteristic of organisms that reproduce sexually. The Austrian monk Gregor Mendel performed thousands of crosses with garden peas at his monastery during the middle of the nineteenth century.

Modification drive A gene drive designed to spread genomic changes and/or genetic payloads throughout a population, thereby modifying the population.

Monogenic Involving or controlled by a single gene.

Mutagenesis Mutagenesis is a process by which the genetic information of an organism is changed, resulting in a mutation. It may occur spontaneously in nature or as a result of exposure to mutagens. It can also be achieved experimentally using laboratory procedures.

Mutagenic chain reaction An approach that employs the CRISPR system to drive a mutation to high frequency in a population, making gene replacement at the population level practical for any species that can be made to accept a transgene in the laboratory, which is based on the CRISPR genome editing system for generating autocatalytic mutations to generate homozygous loss-of-function mutations.

NHEJ Nonhomologous end joining (NHEJ) is a pathway that repairs double-strand breaks in DNA. NHEJ is referred to as "nonhomologous" because the break ends are directly ligated without the need for a homologous template, in contrast to homology-directed repair, which requires a homologous sequence to guide repair.

Nuclease A nuclease is an enzyme capable of cleaving the phosphodiester bonds between nucleotides of nucleic acids. Nucleases variously effect single and double-stranded breaks in their target molecules. In living organisms, they are essential machinery for many aspects of DNA repair.

Off target Off-target genome editing refers to nonspecific and unintended genetic modifications that can arise through the use of engineered nuclease technologies such as CRISPR, TALENs and meganucleases. These tools use different mechanisms to bind a predetermined sequence of DNA target, which they cleave or cut creating a double-stranded chromosomal break [DSB] that summons the cell's DNA repair mechanisms (nonhomologous end joining [NHEJ] and homologous recombination [HR]) and leads to site-specific modifications. If these complexes do not bind at the target, often a result of homologous sequences and/or mismatch tolerance, they will cleave off-target and cause nonspecific genetic modifications. Specifically, off-target effects consist of unintended point mutations, deletions, insertions, inversions and translocations.

Oligonucleotides Oligonucleotides are short DNA or RNA molecules, oligomers that have a wide range of applications in genetic testing, research and forensics.

Ovalbumin Ovalbumin (abbreviated OVA) is the main protein found in egg white, making up approximately 55% of the total protein. Ovalbumin displays sequence and three-dimensional homology to the serpin superfamily, but unlike most serpins, it is not a serine protease inhibitor.

Ovomucoid Ovomucoid is a protein found in egg whites. It is a trypsin inhibitor with three protein domains of the Kazal domain family; it is not the same as ovomucin, another egg white protein. Chicken ovomucoid, also known as Gal d 1, is a known allergen. It is the protein most often causing egg allergy. At least four IgE epitopes have been identified.

PAM The protospacer adjacent motif (PAM) is a short DNA sequence (usually 2–6 base pairs in length) that follows the DNA region targeted for cleavage by the CRISPR system, such as CRISPR-Cas9. The PAM is required for a Cas nuclease to cut.

Parthenogenesis A type of asexual reproduction in which the offspring develops from an unfertilized egg.

Pathogenic Pathogenic means able to cause disease in a person, animal or plant; examples are bacteria, viruses or other microorganisms causing disease.

Payload genes Genes that can be linked to a gene drive to spread a desirable trait throughout a population.

Phenotype Phenotype is the term used in genetics for the composite observable characteristics or traits of an organism, such as height, eye color and blood type. The genetic contribution to the phenotype is called the genotype. Some traits are largely determined by the genotype, while other traits are largely determined by environmental factors.

Polygenic A polygenic trait is one whose phenotype is influenced by more than one gene. Traits that display a continuous distribution, such as height or skin color, are polygenic.

Progeny Offspring of animals or plants.

Prokaryote Also spelled procaryote; any organism that lacks a distinct nucleus and other organelles due to the absence of internal membranes. Bacteria are among the best-known prokaryotic organisms.

Recombinant adeno-associated virus (rAAV) Recombinant adeno-associated virus (rAAV) is a genetically engineered AAV that enables insertion, deletion, or substitution of DNA sequences into live mammalian cells' genomes.

Recombinant DNA technology Recombinant DNA technology is the joining together of DNA molecules from two different species. The recombined DNA molecule is inserted into a host organism to produce new genetic combinations that are of value to science, medicine, agriculture and industry.

Removability The ability to completely remove a gene drive system from a population.

Repeat variable diresidue This is a repeated highly conserved 33–34 amino acid sequence with divergent 12th and 13th amino acids contained in the DNA binding domain. These two positions, referred to as the repeat variable diresidue (RVD). They are highly variable and show a strong correlation with specific nucleotide recognition.

Resistance alleles Alleles that are resistant to a drive system, preventing it from spreading. They can originate from mutations or errors in replication or DNA repair resulting from the gene drive, or they can exist in a population before the release of the gene drive.

Reversibility The ability to replace an existing gene drive system with another system.

RGENs They include CRISPR. RNA-guided endonucleases (RGENs) are novel, programmable genome engineering tools that were developed from bacterial adaptive immune machinery.

RNA Ribonucleic acid (RNA) is an important biological macromolecule that is present in all biological cells. It is composed of ribose and phosphate groups. It has the base uracil which is not found in DNA.

RNA-guided drive Any engineered drive system that utilizes an RNA-guided endonuclease to bias its inheritance and to increase in frequency in a population.

SaCas9 SaCas9 is an endonuclease. SaCas9 is compact and can be packaged in the payload-limited adeno-associated viral (AAV) vector that is commonly used for in vivo gene editing.

SDN Site-directed nuclease technology, SDN-1, produces a double-stranded break in the genome of a plant without the addition of foreign DNA. The template contains one or several small sequence changes in the genomic code that the repair mechanism copies into the plant's genetic material, resulting in a mutation of the target gene.

Segregation and transmission distorters Segregation distorters are alleles that distort normal segregation in their own favor. Sex chromosomal distorters lead to biased sex ratios and the presence of such distorters therefore may induce selection for a change in the mechanism of sex determination.

Selfish genetic elements Selfish genetic elements are genetic segments that can enhance their own transmission at the expense of other genes in the genome, even if this has no positive or a net negative effect on organismal fitness.

Self-limiting drive The are drives that are limited in time and space. This occurs regardless of physical or ecological barriers or migration level.

***Semele* system** *Semele* consists of two components: a toxin expressed in transgenic males that either kills or renders infertile wild-type female recipients and an antidote expressed in females that protects them from the effects of the toxin.

Sequencing Sequencing is the process of determining the nucleic acid sequence characters. For example,—the order of nucleotides in DNA. It includes any method or technology that is used to determine the order of the four bases: adenine, guanine, cytosine and thymine.

Sex-linked meiotic drives These are sex-ratio distorters. They function by biasing the gender ratio of offspring. This can be such that the larger fraction of offspring is a particular gender for example 90% males and 10% females.

Spermatozoa The mature motile male sex cell of an animal by which the ovum is fertilized, typically having a compact head and one long flagella for swimming.

Stem cells A stem cell is a cell with the unique ability to develop into specialized cell types in the body. In the future, they may be used to replace cells and tissues that have been damaged or lost due to disease.

Sterile insect technique A method for temporarily suppressing target populations, whereby overwhelming numbers of mass-produced sterile insects are released to mate with wild-type insects.

Supernumerary Present in excess of the normal or requisite number.

Suppression drive A gene drive designed to reduce or eliminate the population of its target organism. Suppression drives typically work by using non-Mendelian inheritance to spread alleles that cause lethality or sterility or skew the offspring sex ratio, typically toward males.

Switching off Each cell expresses, or turns on, only a fraction of its genes. The rest of the genes are repressed, or turned off. The process of making genes not to be expressed is called switching off.

TAL effector TAL effectors are proteins secreted by bacterial pathogens into plant cells, where they enter the nucleus and activate expression of individual genes.

TALENs Transcription activator-like effector nucleases are restriction enzymes that can be engineered to cut specific sequences of DNA. They are made by fusing a TAL effector DNA-binding domain to a DNA cleavage domain.

T-complex T-complex protein 1 subunit alpha[a] is a protein that in humans is encoded by the TCP1 gene.

T-haplotype A haplotype (haploid genotype) is a group of alleles in an organism that are inherited together from a single parent. The t-haplotype (t) is a male meiotic driver in the house mouse, *Mus musculus*.

Threshold-dependent drives Threshold-dependent drivesare drives which only spread when they are released in a population above a critical frequency. However, under certain conditions, small changes in gene drive fitness could lead to divergent outcomes in spreading behavior.

tracrRNA Trans-activating CRISPR RNA, or "tracer RNA," is a small trans-encoded RNA. In the CRISPR-Cas9 system, the tracrRNA base pairs with the crRNA to form a functional guide RNA (gRNA). Cas9 uses the tracrRNA portion of the guide as a handle, while the crRNA spacer sequence directs the complex to a matching viral sequence.

Trait stacking Gene stacking refers to the introduction of two or more transgenes of agronomic interest in the same plant. Trait stacking for genetically modified (GM) or biotech crops refers to the incorporation of multiple genetic modifications, or traits, in a single variety of a crop.

Trans-acting Trans-acting (trans-regulatory, trans-regulation), in general, means "acting from a different molecule" (i.e., intermolecular). Both the trans-acting gene and the protein/RNA that it encodes are said to "act in trans" on the target gene.

Transgenic Transgenic means that one or more DNA sequences from another species have been introduced by artificial means. Transgenic plants can be made by introducing foreign DNA into a variety of different tissues.

Translocations Translocations generate novel chromosomes. In a translocation, a segment from one chromosome is transferred to a nonhomologous chromosome or to a new site on the same chromosome. Translocations place genes in new linkage relationships and generate chromosomes without normal pairing partners.

Transposable element A transposable element (TE, transposon, or jumping gene) is a DNA sequence that can change its position within a genome, sometimes creating or reversing mutations and altering the cell's genetic identity and genome size. Transposition often results in duplication of the same genetic material.

Transposons A class of genetic elements that can insert themselves into different locations in a genome.

Underdominance Unidirectional mens in one direction. This is the opposite of overdominance. It is the selection against the heterozygote, causing disruptive selection and divergent genotypes. Underdominance exists in situations where the heterozygotic genotype is inferior in fitness to either the dominant or recessive homozygotic genotype.

Unidirectional In unidirectional replication, growth proceeds along both strands in the same direction leading from the origin. Along one of the parental template strands, synthesis of the new complementary strand takes place by the continuous addition of nucleotides to the available 3′ end of the forming strand.

Viral genomes Viral genome means the genetic material of a virus. Viral genomes consist of DNA or RNA only, never both. DNA and RNA molecules can be double-stranded or single-stranded, linear or circular, segmented (composed of multiple pieces of nucleic acid) or nonsegmented. The main function of the virion is to deliver its DNA or RNA genome into the host cell so that the genome can be expressed (transcribed and translated) by the host cell. The viral genome, often with associated basic proteins, is packaged inside a symmetric protein capsid.

Wild type A strain, gene, or characteristic which prevails among individuals in natural conditions, as distinct from an atypical mutant type. Wild type refers to the phenotype of the typical form of a species as it occurs in nature. Originally, the wild type was conceptualized as a product of the standard "normal" allele at a locus, in contrast to that produced by a nonstandard, "mutant" allele.

Wolbachia *Wolbachia* is a genus of intracellular bacteria that infects mainly arthropod species, including a high proportion of insects and also some nematodes. It is one of the most common parasitic microbes and is possibly the most common reproductive parasite in the biosphere.

X chromosome The X chromosome is one of two sex chromosomes. Humans and most mammals have two sex chromosomes, the X and Y. Females have two X chromosomes in their cells, while males have X and Y chromosomes in their cells. Egg cells all contain an X chromosome, while sperm cells contain an X or a Y chromosome.

Xenografting A tissue graft or organ transplant from a donor of a different species from the recipient. Xenotransplantation or heterologous transplant is the transplantation of living cells, tissues or organs from one species to another.

X-linked X-linked recessive inheritance refers to genetic conditions associated with mutations in genes on the X chromosome. A male carrying such a mutation will be affected, because he carries only one X chromosome. A female carrying a mutation in one gene, with a normal gene on the other X chromosome, is generally unaffected.

Y-chromosome Y is normally the sex-determining chromosome in many species, since it is the presence or absence of Y that typically determines the male or female sex of offspring produced in sexual reproduction. In mammals, the Y chromosome contains the gene SRY, which triggers male development.

X-shredder In an X-Y heterogametic species, an X-shredder is a type of gene drive that cleaves the X chromosome at multiple places during meiosis in males, thus destroying it. Because of this, most or all of the viable sperm will contain Y chromosomes, resulting in biased sex ratios in favor of males and, over time, suppression of the population, owing to lack of females.

Y-linked These are traits whose inheritance is linked to the Y chromosome. Y-linked traits never occur in females and occur in all male descendants of an affected male. The concepts of dominant and recessive do not apply to Y-linked traits, as only one allele (on the Y) is ever present in any one (male) individual.

Zinc-binding The binding of zinc finger. It is is found to be distinct from many other DNA-binding proteins that bind DNA through the twofold symmetry of the double helix; instead, zinc fingers are linked linearly in tandem to bind nucleic acid sequences of varying lengths. Zinc fingers often bind to a sequence of DNA known as the GC box.

Zinc finger A zinc finger is a small protein structural motif that is characterized by the coordination of one or more zinc ions (Zn^{2+}) in order to stabilize the fold.

Zinc finger nucleases These are synthetic proteins used for gene targeting. They consist of a DNA-cutting endonuclease domain fused to zinc finger domains engineered to bind a specific DNA sequence. ZFNs are used to introduce insertions or deletions at cut sites in the genomes of living cells.

References

Abbehausen, C. (2019). Zinc Finger Domains as Therapeutic Targets for Metal-Based Compounds—An Update. *Metallomics*. 11, 15–28. https://doi.org/10.1039/C8MT00262B.

Aglawe, S. B., Barbadikar, K. M., Mangrauthia, S. K. and Madhav, M. S. (2018). New Breeding Technique "Genome Editing" for Crop Improvement: Applications, Potentials and Challenges. *Biotechnology*. 8(8), 1–20. https://doi.org/10.1007/s13205-018-1355-3

Ahmad, S. F. and Martins, C. (2019). The Modern View of B Chromosomes under the Impact of High Scale Omics Analyses. *Cells*. 8, 156.

Ahmed, S., Zhang, Y. and Abdullah, M. (2019). Current Status, Challenges, and Future Prospects of Plant Genome Editing in China. *Plant Biotechnology Reports*. 13, 459–472.

Akbari, O. S., Matzen, K. D., Marshall, J. M., Huang, H., Ward, C. M. and Hay, B. A. (2013). A Synthetic Gene Drive System for Local, Reversible Modification and Suppression of Insect Populations. *Current Biology*. 23(8), 671–677.

Akbari, O. S., Papathanos, P. A., Sandler, J. E., Kennedy, K. and Hay, B. A. (2014). Identification of Germline Transcriptional Regulatory Elements in *Aedes aegypti*. *Scientific Reports*. 4, 3954.

Ali, Z., Abulfaraj, A., Idris, A., Ali, S., Tashkandi, M. and Mahfouz, M. M. (2015). CRISPR/Cas9-Mediated Viral Interference in Plants. *Genome Biology*. 16, 238.

Alphey, L. (2014). Genetic Control of Mosquitoes. *Annual Review of Entomology.* 59, 205–224.

Ant, T. H., Herd, C., Louis, F., Failloux, A. B. and Sinkins, S. P. (2020). *Wolbachia* Transinfections in *Culex quinquefasciatus* Generate Cytoplasmic Incompatibility. *Insect Molecular Biology*. 29(1), 1–8. https://doi.org/10.1111/imb.12604

Aryan, A., Anderson, M. A., Myles, K. M. and Adelman, Z. N. (2013a). TALEN-Based Gene Disruption in the Dengue Vector *Aedes aegypti*. *PloS One*. 8(3), e60082. https://doi.org/10.1371/journal.pone.0060082

Aryan, A., Anderson, M. A., Myles, K. M. and Adelman, Z. N. (2013b). Germline Excision of Transgenes in *Aedes aegypti* by Homing Endonucleases. *Scientific Reports*. 3, 1603. https://doi.org/10.1038/srep01603

Asimakis, E. D., Doudoumis, V., Hadapad, A. B., Hire, R. S., Batargias, C., Niu, C., Khan, M., Bourtzis, K. and Tsiamis, G. (2019). Detection and Characterization of Bacterial Endosymbionts in Southeast Asian Tephritid Fruit Fly Populations. *BMC Microbiology*. 19, 290.

Asselin, A. K., Villegas-Ospina, S., Hoffmann, A. A., Brownlie, J. C. and Johnson, K. N. (2019). Contrasting Patterns of Virus Protection and Functional Incompatibility Genes in Two Conspecific *Wolbachia* Strains from *Drosophila pandora*. *Applied and Environmental Microbiology*. 85(5), e02290.

Au, R. (2015). From Genetic Engineering to Genome Engineering: What Impact Has It Made on Science and Society? *Advances in Genetic Engineering & Biotechnology*. 2, 1–8.

Auerbach, C., Robson, J. M., Carr, J. G. (1947). Chemical Production of Mutations. *Science*. 105, 243–247.

Autralian Academy of Sciences (AAS). (2017). *Discussion Paper Synthetic Gene Drives in Australia: Implications of Emerging Technologies*. www.science.org.au/gene-drives; www.science.org.au/support/analysis/reports/synthetic-gene-drives-australia-implications-emerging-technologies

Avanesyan, A., Jaffe, B. D. and Guédot, C. (2017). Isolating Spermathecae and Determining Mating Status of *Drosophila suzukii*: A Protocol for Tissue Dissection and Its Applications. *Insects*. 8(1), 1–12. https://doi.org/10.3390/insects8010032

Backus, G. A. and Delborne, J. A. (2019). Threshold-Dependent Gene Drives in the Wild: Spread, Controllability, and Ecological Uncertainty. *BioScience*. 69(11), 900–907. https://doi.org/10.1093/biosci/biz098

Bahrami, A. and Najafi, A. (2019). Synthetic Animal: Trends in Animal Breeding and Genetics. *Insights in Biology and Medicine*. 3, 007–025. https://doi.org/10.29328/journal.ibm.1001015

Baker, M. (2012). Gene-Editing Nucleases. *Nature Methods*. 9(1), 23–26.

Barrangou, R. (2014). RNA Events. Cas9 Targeting and the CRISPR Revolution. *Science*. 16: 344(6185):707–708. https://doi.org/10.1126/science.1252964. PMID: 24833384.

Barrangou, R. and Horvath, P. (2012). CRISPR: New Horizons in Phage Resistance and Strain Identification. *Annual Review of Food Science and Technology*. 3(1), 143–162.

Barrangou, R., Fremaux, C., Deveau, H., Richards, M., Boyaval, P. and Moineau, S. (2007). CRISPR Provides Acquired Resistance against Viruses in Prokaryotes. *Science*. 315(5819), 1709–1712.

Barrangou, R. and Horvath, P. (2017). A Decade of Discovery: CRISPR Functions and Applications. *Nature Microbiology*. 2, 17092. https://doi.org/10.1038/nmicrobiol.2017.92

Barrangou, R. and Marraffini, L. A. (2014). CRISPR-Cas Systems: Prokaryotes Upgrade to Adaptive Immunity. *Molecular Cell*. 54(2), 234–244.

Bartlett, C. and Root, E. (2015). A TALE of Two Nucleases: Using TALENs to Edit the Genome of *C. elegans*. *Bioengineering Senior Theses*. 27. https://scholarcommons.scu.edu/bioe_senior/27

Bauer, D. E., Kamran, S. C., Lessard, S., Xu, J., Fujiwara, Y. and Lin, C. (2013). An Erythroid Enhancer of BCL11A Subject to Genetic Variation Determines Fetal Hemoglobin Level. *Science*. 342, 253–257.

Bauman, S., Yan, Y., Finnigan, G. C., Brossard, D., Belluck, P., Gould, F., Paques, F. et al. (2018). Sheep and Goat Genome Engineering: From Random Transgenesis to the CRISPR Era. *Science*. 8(1), 23–25. https://doi.org/10.1016/j.synbio.2018.09.004

Beaghton, A. K., Hammond, A., Nolan, T., Crisanti, A. and Burt, A. (2019). Gene Drive for Population Genetic Control: Non-Functional Resistance and Parental Effects. *Proceedings of the Society for Experimental Biology and Medicine*. 286(1914), 1586.

Beckmann, J. F., Bonneau, M., Chen, H., Hochstrasser, M., Poinsot, D., Merçot, H., Weill, M., Sicard, M. and Charlat, S. (2019). The Toxin-Antidote Model of Cytoplasmic Incompatibility: Genetics and Evolutionary Implications. *Trends in Genetics: TIG*. 35(3), 175–185. https://doi.org/10.1016/j.tig.2018.12.004

Beckmann, J. F., Ronau, J. A. and Hochstrasser, M. (2017). A *Wolbachia* Deubiquitylating Enzyme Induces Cytoplasmic Incompatibility. *Nature Microbiology*. 2(5), 1–7. https://doi.org/10.1038/nmicrobiol.2017.7

Beckmann, J. F., Sharma, G. D., Mendez, L., Chen, H. and Hochstrasser, M. (2019). The *Wolbachia* Cytoplasmic Incompatibility Enzyme CIDB Targets Nuclear Import and Protamine-Histone Exchange Factors. *eLife*. 8, 1–23. https://doi.org/10.7554/eLife.50026

Benetta, D. E., Akbari, O. S. and Ferree, P. M. (2019). Sequence Expression of Supernumerary B Chromosomes: Function or Fluff? *Genes (Basel)*. 8: 10(2), 123. https://doi.org/10.3390/genes10020123. PMID: 30744010; PMCID: PMC6409846

Beumer, K. J., Trautman, J. K., Christian, M., Dahlem, T. J., Lake, C. M., Hawley, R. S., Grunwald, D. J., Voytas, D. F. and Carroll, D. (2013). Comparing Zinc Finger Nucleases and Transcription Activator-Like Effector Nucleases for Gene Targeting in Drosophila. *G3: Genes, Genomes, Genetics*. 3(10), 1717–1725. https://doi.org/10.1534/g3.113.007260

Bevacqua, R. J., Canel, N. G., Hiriart, M. I., Sipowicz, P., Rozenblum, G. T., Vitullo, A. et al. (2013). Simple Gene Transfer Technique Based on I-SceImeganuclease and Cytoplasmic Injection in IVF Bovine Embryos. *Theriogenology*. 80, 104–113.e29. https://doi.org/10.1016/j.theriogenology.2013.03.017

Bi, J. and Wang, Y. F. (2019). The Effect of the Endosymbiont *Wolbachia* on the Behavior of Insect Hosts. *Insect Science*. 1–13. https://doi.org/10.1111/1744-7917.12731

Bloom, K., Mussolino, C. and Arbuthnot, P. (2015). Transcription Activator-Like Effector (TALE) Nucleases and Repressor TALEs for Antiviral Gene Therapy. *Current Stem Cell Reports*. 1, 1–8. https://doi.org/10.1007/s40778-014-0008-7

Boglioli, E. and Richard, M. (2015). *Rewriting the Book of Life: A New Era in Precision Gene Editing*. www.bcg.com/publications/2015/rewriting-book-of-life-new-era-precision-gene-editing

Boissel, S., Jarjour, J., Astrakhan, A., Adey, A., Gouble, A., Duchateau, P., Shendure, J., Stoddard, B. L., Certo, M. T., Baker, D. and Scharenberg, A. M. (2014). MegaTALs: A Rare-Cleaving Nuclease Architecture for Therapeutic Genome Engineering. *Nucleic Acids Research*. 42(4), 2591–2601. https://doi.org/10.1093/nar/gkt1224

Bolotin, A., Quinquis, B., Sorokin, A. and Ehrlich, S. D. (2005). Clustered Regularly Interspaced Short Palindrome Repeats (CRISPRs) Have Spacers of Extrachromosomal Origin. *Microbiology (Reading)*. 151(Pt 8), 2551–2561. https://doi.org/10.1099/mic.0.28048-0

Bonawitz, N. D., Ainley, W. M., Itaya, A. et al. (2019). Zinc Finger Nuclease-Mediated Targeting of Multiple Transgenes to an Endogenous Soybean Genomic Locus via Non-Homologous End Joining. *Plant Biotechnology Journal*. 17(4), 750–761. https://doi.org/10.1111/pbi.13012

Borsenberger, V., Onésime, D., Lestrade, D., Rigouin, C., Neuvéglise, C., Daboussi, F. and Bordes, F. (2018). Multiple Parameters Drive the Efficiency of CRISPR/Cas9-Induced Gene Modifications in *Yarrowia lipolytica*. *Journal of Molecular Biology*. 430(21), 4293–4306. https://doi.org/10.1016/j.jmb.2018.08.024

Brandt, R. and Bucking, E. (2019). *Gene Drives Report*. https://www.ncbi.nlm.nih.gov/pmc/articles/PMC6939923/

Bridgeman, B., Morgan-Richards, M., Wheeler, D. and Trewick, S. A. (2018). First Detection of *Wolbachia* in the New Zealand Biota. *PLoS One*. 13(4), 9–11. https://doi.org/10.1371/journal.pone.0195517

Brossard, D., Belluck, P., Gould, F. and Wirz, C. D. (2019). Promises and Perils of Gene Drives: Navigating the Communication of Complex, Post-Normal Science. *Proceedings of the National Academy of Sciences*. 116(16), 7692–7697.

Brossarda, D., Belluckc, P., Gouldde, F. and Christopher, D.2018. Promises and Perils of Gene Drives: Navigating the Communication of Complex, Post-Normal Science. *Proceedings of the National Academy of Sciences*. 116(16), 7692–7697. https://doi.org/10.1073/pnas.1805874115

Brouns, S. J. J., Jore, M. M., Lundgren, M., Westra, E. R., Slijkhuis, R. J. H., Snijders, A. P. L., Dickman, M. J., Makarova, K. S., Koonin, E. V. and van der Oost, J. (2008). Small CRISPR RNAs Guide Antiviral Defense in Prokaryotes. *Science*. 321(5891), 960–964. https://doi.org/10.1126/science.1159689

Bruegmann, T., Deecke, K. and Fladung, M. (2019). Evaluating the Efficiency of gRNAs in CRISPR/Cas9 Mediated Genome Editing in Poplars. *International Journal of Molecular Sciences*. 20(15), 3623. https://doi.org/10.3390/ijms20153623

Buchman, A., Marshall, J. M., Ostrovski, D., Yang, T. and Akbari, O. S. (2018). Synthetically Engineered Medea Gene Drive System in the Worldwide Crop Pest *Drosophila suzukii*. *Proceedings of the National Academy of Sciences of the United States of America*. 115(18), 4725–4730. https://doi.org/10.1073/pnas.1713139115

Bull, J. J., Remien, C. H., Gomulkiewicz, R. and Krone, S. M. (2019). Spatial Structure Undermines Parasite Suppression by Gene Drive Cargo. *PeerJ*. 7, e7921. https://doi.org/10.7717/peerj.7921

Burt, A. (2003). Site-Specific Selfish Genes as Tools for the Control and Genetic Engineering of Natural Populations. *Proceedings of the Royal Society of London*. B(270), 20022319. https://doi.org/10.1098/rspb.2002.2319

Burt, A. and Crisanti, A. (2018). Gene Drive: Evolved and Synthetic. *ACS Chemical Biology*. 13(2), 343–346. https://doi.org/10.1021/acschembio.7b01031

Cao, J., Wu, L., Zhang, S., Lu, M., Cheung, W. K. C., Cai, W., Gale, M., Xu, Q. and Yan, Q. (2016). An Easy and Efficient Inducible CRISPR/Cas9 Platform with Improved Specificity for Multiple Gene Targeting. *Nucleic Acids Research*. 44(19), e149. https://doi.org/10.1093/nar/gkw660

Caputo, A. T., Alonzi, D. S., Kiappes, J. L., Struwe, W. B., Cross, A., Basu, S., Darlot, B. and Roversi, P. (2018). *Dengue and Zika: Control and Antiviral Treatment Strategies*. Singapore: Springer, volume 1062, 265–276.

Caragata, E. P. and Moreira, L. A. (2017). *Using an Endosymbiont to Control Mosquito-Transmitted Disease. Arthropod Vector: Controller of Disease Transmission*. Amsterdam: Elsevier Inc., volume 1. https://doi.org/10.1016/B978-0-12-805350-8.00007-6

Caragata, E. P., Rocha, M. N., Pereira, T. N., Mansur, S. B., Dutra, H. L. C. and Moreira, L. A. (2019). Pathogen Blocking in *Wolbachia*-Infected *Aedes aegypti* Is Not Affected by Zika and Dengue Virus Co-Infection. *PLoS Neglected Tropical Diseases*. 13(5), 1–26. https://doi.org/10.1371/journal.pntd.0007443

Carrington, L. B., Tran, B. C. N., Le, N. T. H., Luong, T. T. H., Nguyen, T. T., Nguyen, P. T., Nguyen, C. V. V. et al. (2017). Field- and Clinically Derived Estimates of *Wolbachia*-Mediated Blocking of Dengue Virus Transmission Potential in *Aedes aegypti* Mosquitoes. *Proceedings of the National Academy of Sciences of the United States of America*. 115(2), 361–366. https://doi.org/10.1073/pnas.1715788115

Carroll, D. (2011). Genome Engineering with Zinc-Finger Nucleases. *Genetics*. 188(4), 773–782. https://doi.org/10.1534/genetics.111.131433

Carroll, D. (2014). Genome Engineering with Targetable Nucleases. *Annual Review of Biochemistry*. 83, 409–439.

Carroll, D. (2017). Genome Editing: Past, Present, and Future. *Yale Journal of Biology and Medicine*. 90(4), 653–659.

Carter, S. R. F. R. (2016). *Policy and Regulatory Issues for Gene Drives in Insects*. San Diego: J. Craig Venter Institute, 1–21.

Carvajal, T. M., Hashimoto, K., Harnandika, R. K., Amalin, D. M. and Watanabe, K. (2019). Detection of *Wolbachia* in Field-Collected *Aedes aegypti* Mosquitoes in Metropolitan Manila, Philippines. *Parasites and Vectors*. 12(1), 361. https://doi.org/10.1186/s13071-019-3629-y

Cassandri, M., Smirnov, A., Novelli, F., Pitolli, C., Agostini, M., Malewicz, M. and Raschellà, G. (2017). Zinc-Finger Proteins in Health and Disease. *Cell Death Discovery*. https://doi.org/10.1038/cddiscovery.2017.71

Cathomen, T. and Joung, J. K. (2008). Zinc-Finger Nucleases: The Next Generation Emerges. *Molecular Therapy*. 16(7), 1200–1207. https://doi.org/10.1038/mt.2008.114

Champer, J., Buchman, A. and Akbari, O. S. (2016). Cheating Evolution: Engineering Gene Drives to Manipulate the Fate of Wild Populations. *Nature Reviews Genetics*. 17(3), 146–159. https://doi.org/10.1038/nrg.2015.34

Champer, J., Reeves, R., Oh, S. Y., Liu, C., Liu, X., Clark, A. G. and Messer, P. W. (2016). Novel CRISPR/ Cas9 Gene Drive Constructs Reveal Insights into Mechanisms of Resistance Allele Formation and Drive Efficiency in Genetically Diverse Population. *PLoS Genetics*. 13(7).

Champer, J., Reeves, R., Oh, S. Y., Liu, C., Liu, J., Clark, A. G. and Messer, P. W. (2017). Novel CRISPR/Cas9 Gene Drive Constructs Reveal Insights into Mechanisms of Resistance Allele Formation and Drive Efficiency in Genetically Diverse Populations. *PLoS Genetics*. 13(7), 1–18. https://doi.org/10.1371/journal.pgen.1006796

Chandrasegaran, S. and Carroll, D. (2016). Origins of Programmable Nucleases for Genome Engineering. *Journal of Molecular Biology*. 428, 963–989.

Chang, S., Sung, P. S., Lee, J., Park, J., Shin, E. C. and Choi, C. (2016). Prolonged Silencing of Diacylglycerol Acyltransferase-1 Induces a Dedifferentiated Phenotype in Human Liver Cells. *Journal of Cellular and Molecular Medicine*. 20(1), 38–47. https://doi.org/10.1111/jcmm.12685

Chapman, J. E., Gillum, D. and Kiani, S. (2017). Approaches to Reduce CRISPR Off-Target Effects for Safer Genome Editing. *Applied Biosafety*. 22(1), 7–13. https://doi.org/10.1177/1535676017694148

Charlesworth, C. T. (2019). *Identification of Pre-Existing Adaptive Immunity to Cas9*. https://www.biorxiv.org/content/10.1101/243345v1

Chaverra-Rodriguez, D., Macias, V. M., Hughes, G. L., Pujhari, S., Suzuki, Y., Peterson, D. R., Kim, D., McKeand, S. and Rasgon, J. L. (2018). Targeted Delivery of CRISPR-Cas9 Ribonucleoprotein into Arthropod Ovaries for Heritable Germline Gene Editing. *Nature Communications*. 9(1), 1–11. https://doi.org/10.1038/s41467-018-05425-9

Chegeni, T. N. and Fakhar, M. (2019). Promising Role of *Wolbachia* as Anti-Parasitic Drug Target and Eco-Friendly Biocontrol Agent. *Recent Patents on Anti-Infective Drug Discovery*. 14(1), 69–79. https://doi.org/10.2174/1574891x14666190211162403

Chen, C. H., Huang, H., Ward, C. M., Su, J. T., Schaeffer, L. V., Guo, M. and Hay, B. A. (2007). A Synthetic Maternal-Effect Selfish Genetic Element Drives Population Replacement in Drosophila. *Science*. 316(5824), 597–600.

Chen, H., Ronau, J. A., Beckmann, J. F. and Hochstrasser, M. (2019). A *Wolbachia* Nuclease and Its Binding Partner Provide a Distinct Mechanism for Cytoplasmic Incompatibility. *Proceedings of the National Academy of Sciences of the United States of America*, 116(44), 22314–22321. https://doi.org/10.1073/pnas.1914571116

Chen, L., Wang, G., Zhu, Y. N., Xiang, H. and Wang, W. (2016). Advances and Perspectives in the Application of CRISPR/Cas9 in Insects. *Dong wu xue yan jiu = Zoological Research*. 37(4), 220–228. https://doi.org/10.13918/j.issn.2095-8137.2016.4.220

Chen, S. J. (2019). Minimizing Off-Target Effects in CRISPR-Cas9 Genome Editing. *Cell Biology and Toxicology*. 35(5), 399–401. https://doi.org/10.1007/s10565-019-09486-4

Chiu, J. C., Jiang, X., Zhao, L., Hamm, C. A., Cridland, J. M., Saelao, P., Hamby, K. A., Lee, E. K., Kwok, R. S., Zhang, G., et al. (2013). Genome of *Drosophila suzukii*, the Spotted Wing Drosophila. *G3 (Bethesda)*, 3, 2257–2271.

Cho, W. S. S., Kim, J. M. and Kim, J. S. (2013). Targeted Genome Engineering in Human Cells with the Cas9 RNA-Guided Endonuclease. *Nature Biotechnology*. 31, 230–232.

Chouin-Carneiro, T., Ant, T. H., Herd, C., Louis, F., Failloux, A. B. and Sinkins, S. P. (2020). *Wolbachia* Strain wAlbA Blocks Zika Virus Transmission in *Aedes aegypti*. *Medical and Veterinary Entomology*. 34(1), 116–119. https://doi.org/10.1111/mve.12384

Chrostek, E. and Gerth, M. (2019). Is *Anopheles gambiae* a Natural Host of *Wolbachia*? *MBio*. 10(3), 1–10. https://doi.org/10.1128/mBio.00784-19

Cong, L., Ran, F. A., Cox, D., Lin, S. et al., 2013. Multiplex Genome Engineering Using CRISPR/Cas Systems. *Science*. 339, 819–823.

Congressional Research Service. (2018). *Advanced Gene Editing: CRISPR-Cas9*. https://fas.org/sgp/crs/misc/R44824.pdf

Cornell, C. (2020). *CRISPR-Cas3 Innovation Holds Promise for Disease Cures, Advancing Science*. https://www.sciencedaily.com/releases/2019/04/190411172519.htm

Courtier-Orgogozo, V., Morizot, B. and Boëte, C. (2017). Agricultural Pest Control with CRISPR-Based Gene Drive: Time for Public Debate: Should We Use Gene Drive for Pest Control? *EMBO Reports*. 18(6), 878–880. https://doi.org/10.15252/embr.201744205

Crowther, M. D., Dolton, G. and Legut, M. (2020). Genome-wide CRISPR—Cas9 Screening Reveals Ubiquitous T Cell Cancer Targeting via the Monomorphic MHC Class I-Related Protein MR1. *Nature Immunology*. 21, 178–185. https://doi.org/10.1038/s41590-019-0578-8

Daboussi, F., Stoddard, T. J. and Zhang, F. (2015). Engineering Meganuclease for Precise Plant Genome Modification. *Advances in New Technology for Targeted Modification of Plant Genomes*. New York, NY: Springer, 21–38. https://doi.org/10.1007/978-1-4939-2556-8_2

Davidson, B., Davidson, A. R., Lu, W., Stanley, S. Y., Wang, J., Mejdani, M., Trost, C. N., Hicks, B. T., Lee, J. and Sontheimer, E. J. (2020). Anti-CRISPRs: Protein Inhibitors of CRISPR. *Cas Systems*. 1–24.

Davis, S., Bax, N. and Grewe, P. (2001). Engineered Underdominance Allows Efficient and Economical Introgression of Traits into Pest Populations. *Journal of Theoretical Biology*. 212, 83–98.

Deborah, M., Thurtle, S. and TeWen, Lo. (2018). Molecular Biology at the Cutting Edge: A Review on CRISPR/CAS9 Gene Editing for Undergraduates. *Biochemistry and Molecular Biology Education*. 46(2), 195–205. https://doi.org/10.1002/bmb.21108

DeFrancesco, L. (2011). Move Over ZFNs: A New Technology for Genome Editing May Put the Zinc Finger Nuclease Franchise Out of Business, Some Believe. Not so Fast, Say the Finger People. *Nature Biotechnology*. 29(8), 681+. Gale OneFile: Health and Medicine [Accessed April 21, 2020].

del Amo, L. V., Bishop, A. L., Sánchez, C., H. M., Bennett, J. B., Feng, X., Marshall, J. M., Bier, E. and Gantz, V. M. (2019). Split-Gene Drive System Provides Flexible Application for Safe Laboratory Investigation and Potential Field Deployment Split-Gene Drive System Provides Flexible Application for Safe Laboratory Investigation and Potential Field Deployment. *bioRxiv*. https://doi.org/10.1101/684597

Delhove, J. M. K. M. and Qasim, W. (2017). Genome-Edited T Cell Therapies. *Current Stem Cell Reports*. 3(2), 124–136. https://doi.org/10.1007/s40778-017-0077-5

Deltcheva, E., Chylinski, K., Sharma, C. M., Gonzales, K., Chao, Y., Pirzada, Z. A., Eckert, M. R., Vogel, J. and Charpentier, E. (2011). CRISPR RNA Maturation by Trans-Encoded Small RNA and Host Factor RNase III. *Nature*. 471, 602–607.

Dhole, S., Lloyd, A. L. and Japa, G. (2020). Gene Drive Dynamics in Natural Populations: The Importance of Density-Dependence, Space and Sex. *Annual Review of Ecology, Evolution, and Systematics*. 51(1), 505–531.

Dionysopoulou, N. K., Papanastasiou, S. A., Kyritsis, G. A. and Papadopoulos, N. T. (2020). Effect of Host Fruit, Temperature and *Wolbachia* Infection on Survival and Development of *Ceratitis capitata* Immature Stages. *PLoS One*. 15(3), 1–19. https://doi.org/10.1371/journal.pone.0229727

Doench, J. G., Fusi, N., Sullender, M., Hegde, M., Vaimberg, E. W., Donovan, K. F., Smith, I., Tothova, Z., Wilen, C., Orchard, R., Virgin, H. W., Listgarten, J. and Root, D. E. (2016). Optimized sgRNA Design to Maximize Activity and Minimize Off-Target Effects of CRISPR—Cas9. *Nature Biotechnology*. 34, 1–12. https://doi.org/10.1080/13102818.2017.1406823

Doudna, J. A. and Charpentier, E. (2014). Genome Editing. The New Frontier of Genome Engineering with CRISPR-Cas9. *Science*. 346(6213), 1258096. https://doi.org/10.1126/science.1258096. PMID: 25430774

Drury, D., Dapper, A., Siniard, D., Zentner, G. and Wade, M. (2017). CRISPR/Cas9 Gene Drives in Genetically Variable and Nonrandomly Mating Wild Populations. *Science Advances*. 3(5), e1601910.

Durai, S., Mani, M., Kandavelou, K., Wu, J., Porteus, M. H. and Chandrasegaran, S. (2005). Zinc Finger Nucleases: Custom-Designed Molecular Scissors for Genome Engineering of Plant and Mammalian Cells. *Nucleic Acids Research*. 33(18), 5978–5990. https://doi.org/10.1093/nar/gki912

Eckhoff, P. A., Wenger, E. A., Godfray, H. C. and Burt, A. (2017). Impact of Mosquito Gene Drive on Malaria Elimination in a Computational Model with Explicit Spatial and Temporal Dynamics. *Proceedings of the National Academy of Sciences of the United States of America*. 114, E255–E264. https://doi.org/10.1073/pnas.16110 64114

Edgell, D. R. and Stoddard, B. L. (2011). Tapping Natural Reservoirs of Homing Endonucleases for Targeted Gene Modification. *Proceedings of the National Academy of Sciences of the United States of America*. 108, 13077–13082.

Edgington, M. P., Harvey-Samuel, T. and Alphey, L. (2020). Population-Level Multiplexing, a Promising Strategy to Manage the Evolution of Resistance against Gene Drives Targeting a Neutral Locus. *Evolutionary Applications*. https://doi.org/10.1111/eva.12945.

Egli, D., Zuccaro, M. V., Kosicki, M., Church, G. M., Bradley, A. and Jasin, M. (2018). Inter-Homologue Repair in Fertilized Human Eggs? *Nature*. 560(7717), E5–E7. https://doi.org/10.1038/s41586-018

Eidgenossenschaft, S., Svizzera, C. and Confederation, S. (2018). *Gene Drives*, 1–10. https://www.ccne-ethique.fr/sites/default/files/ekah_bericht_gene_drives_en.pdf

Emerson, C., James, S., Littler, K. and Randazzo, F. (2017). Principles for Gene Drive Research. *Science*. 358(6367), 1135–1136. https://doi.org/10.1126/science.aap9026

Enzmann, B. (2018). *Synthego Full Stack Genome Engineering*. www.synthego.com/blog/gene-drive-crispr [Accessed April 25, 2020].

Esvelt, K. (2017). *Gene Drive Technology: The Thing to Fear Is Fear Itself*. https://hdl.handle.net/1920/11337

Esvelt, K. (2019). *Gene Drives: In Sculpting Evolution*. https://rivm.openrepository.com/bitstream/handle/10029/596003/20150196.pdf?sequence=3andisAllowed=y%0Awww.sculptingevolution.org/genedrives

Esvelt, M. K. and Gemmell, J. N. (2017). Conservation Demands Safe Gene Drive. *PLoS Biology*. 15(11), e2003850. https://doi.org/10.1371/journal.pbio.2003850

Esvelt, M. K., Smidler, A. L., Catteruccia, F. and Church, G. M. (2014). Emerging Technology: Concerning RNA-Guided Gene Drives for the Alteration of Wild Populations. *eLife*. https://doi.org/10.7554/eLife.03401.001.

Evans, S. W. and Palmer, M. J. (2018). Anomaly Handling and the Politics of Gene Drives. *Journal of Responsible Innovation*. 5, S223–S242. https://doi.org/10.1080/23299460.2017.1407911

Faure, G., Shmakov, S. A., Makarova, K. S., Wolf, Y. I., Crawley, A. B., Barrangou, R. and Koonin, E. V. (2019). Comparative Genomics and Evolution of Trans-Activating RNAs in Class 2 CRISPR-Cas Systems. *RNA Biology*. 16(4), 435–448. https://doi.org/10.1080/15476286.2018.1493331

Fears, R. and ter Meulen, V. (2018). Assessing Security Implications of Genome Editing: Emerging Points from an International Workshop. *Frontiers in Bioengineering and Biotechnology*. 6, 34.

Feng, Y., Zhang, S. and Huang, X. (2014). A Robust TALENs System for Highly Efficient Mammalian Genome Editing. *Scientific Reports*. 4, 3632. https://doi.org/10.1038/srep03632

Fleming, A., Abdalla, E. A., Maltecca, C. and Baes, C. F. (2018). Invited Review: Reproductive and Genomic Technologies to Optimize Breeding Strategies for Genetic Progress in Dairy Cattle. *Archives Animal Breeding*. 61(1), 43–57. https://doi.org/10.5194/aab-61-43-2018

Flisikowska, T., Kind, A. and Schnieke, A. (2014). Genetically Modified Pigs to Model Human Diseases. *Journal of Applied Genetics*. 55, 53–64.

Flores, H. A. and O'Neill, S. L. (2018). Controlling Vector-Borne Diseases by Releasing Modified Mosquitoes. *Nature Reviews Microbiology*. 16(8), 508–518. https://doi.org/10.1038/s41579-018-0025-0

Foley, J. A., Ramankutty, N., Brauman, K. A., Cassidy, E. S., Gerber, J. S. and Johnston, M. (2011). Solutions for a Cultivated Planet. *Nature*. 478, 337–342.

Foss, V. D., Hochstresses, L. M. and Wilson, C. R. (2018). Clinical Applications of CRISPR Based Genome Editing and Diagonistics. *Transfusion*. https://doi.org/101111/trf.15126

Frank, H. and Charpentier, E. (2016). CRISPR-Cas Biology, Mechanisms and Relevance. *Philosophical Transactions of the Royal Society*. 8.

Fraser, M. J. Jr. (2012). Insect Transgenesis: Current Applications and Future Prospects. *Annual Review of Entomology*. 57, 267–289. https://doi.org/10.1146/annurev.ento.54.110807.090545. PMID: 22149266.

Fraser, J. E., De Bruyne, J. T., Iturbe-Ormaetxe, I., Stepnell, J., Burns, R. L., Flores, H. A. and O'Neill, S. L. (2017). Novel *Wolbachia*-Transinfected *Aedes aegypti* Mosquitoes Possess Diverse Fitness and Vector Competence Phenotypes. *PLoS Pathogens*. 13(12), 1–19. https://doi.org/10.1371/journal.ppat.1006751

Frieß, J. L., von Gleich, A. and Giese, B. (2019). Gene Drives as a New Quality in GMO Releases—A Comparative Technology Characterization. *PeerJ*. 7, e6793. https://doi.org/10.7717/peerj.6793

Friedrichs, S., Takasu, Y., Kearns, P., Dagallier, B., Oshima, R., Schofield, J. and Moreddu, C. (2019). Meeting Report of the OECD Conference on Genome Editing: Applications in Agriculture-Implications for Health, Environment and Regulation. *Transgenic Research*. 28, 419–463. https://doi.org/10.1007/s11248-019-00154-1

Fu, B. X., Wainberg, M., Kundaje, A. and Fire, A. Z. (2017). High-Throughput Characterization of Cascade Type I-E CRISPR Guide Efficacy Reveals Unexpected PAM Diversity and Target Sequence Preferences. *Genetics*. 206, 1727–1738.

Funkhouser-Jones, L. J., van Opstal, E. J., Sharma, A. and Bordenstein, S. R. (2018). The Maternal Effect Gene Wds Controls *Wolbachia* Titer in Nasonia. *Current Biology*. 28(11), 1692–1702.e6. https://doi.org/10.1016/j.cub.2018.04.010

Gaj, T., Gersbach, C. A. and Iii, C. F. B. (2013). ZFN, TALEN, and CRISPR/Cas-Based Methods for Genome Engineering. *Trends in Biotechnology*. 31(7), 397–405. https://doi.org/10.1016/j.tibtech.2013.04.004

Gaj, T., Sirk, S. J., Shui, S. and Liu, J. (2016). *Genome-Editing Technologies: Principles and Applications*. New York: Cold Spring Harbor Laboratory Press.

Galetto, R., Duchateau, P. and Pâques, F. (2009). Targeted Approaches for Gene Therapy and the Emergence of Engineered Meganucleases. *Expert Opinion on Biological Therapy*. 9(10), 1289–1303, https://doi.org/10.1517/14712590903213669.

Gantz, V. M. and Bier, E. (2016). The Dawn of Active Genetics. *BioEssays*. 38(1), 50–63.

Gao, C. (2018). The Future of CRISPR Technologies in Agriculture. *Nature Reviews Molecular Cell Biology*. 19, 1–2.

Garneau, J. E., Dupuis, M. E. and Villion, M. (2010). The CRISPR/Cas Bacterial Immune System Cleaves Bacteriophage and Plasmid DNA. *Nature*. 468(7320), 67–71.

Georgidis, C. and Quasim, W. (2017). Emerging Applications of Gene Edited T Cells for the Treatment of Leukaemia. *Expert Review of Hematology*. 10(9), 753–755.1080/17474086.2017.1350575

Gherardin, N. A. (2016). Diversity of T Cells Restricted by the MHC Class I-Related Molecule MR1 Facilitates Differential Antigen Recognition. *Immunity*. 44, 32–45.

Gherardin, N. A. (2017). Enumeration, Functional Responses and Cytotoxic Capacity of MAIT Cells in Newly Diagnosed and Relapsed Multiple Myeloma. *Scientific Reports*. 8, 4159.

Giese, B., Frieß, L. J., Barton, H. N., Messer, W. P., Débarre, F., Schetelig, F. M., Windbichler, N., Meimberg, H. and Boëte, C. (2019). *Gene Drives: Dynamics and Regulatory Matters*. A Report from the Workshop Evaluation of Spatial and Temporal Control of Gene Drives. Vienna: MIT Media Lab; Cambridge, MA: Massachusetts Institute of Technology, United States of America, Department of Anatomy; Dunedin, New Zealand: University of Otago, April 4–5.

Glover, B., Akinbo, O., Savadogo, M. Timpo, S., Lemgo, G., Sinebo, W., Akile, S., Obukosia, S., Ouedraogo, J., Ndomondo-Sigonda, M., Koch, M., Makinde, D. and Ambal, A. (2018). Strengthening Regulatory Capacity for Gene Drives in Africa: Leveraging NEPAD's Experience in Establishing Regulatory Systems for Medicines and GM Crops in Africa. *BMC Proceedings*. 12, 11. https://doi.org/10.1186/s12919-018-0108-y

Godfray, H. C. J., North, A and Burt, A. (2017). How Driving Endonuclease Genes Can Be Used to Combat Pests and Disease Vectors. *BMC Biology*. 15(1), 81.

Gokcezade, J., Sienski, G. and Duchek, P. (2014). Efficient CRISPR/Cas9 Plasmids for Rapid and Versatile Genome Editing in Drosophila. *G3: Genes, Genomes, Genetics*. 2279–2282. https://doi.org/10.1534/g3.114.014126

Goold, H. D., Wright, P. and Hailstones, D. (2018). Emerging Opportunities for Synthetic Biology in Agriculture. *Genes*. 9(7). https://doi.org/10.3390/genes9070341

Gould, F., Huang, Y., Legros, M. and Lloyd, A. L. (2008). A Killer-Rescue System for Selflimiting Gene Drive of Anti-Pathogen Constructs. *Proceedings of the Royal Society B: Biological Sciences*. 275, 2823–2829.

Greiner, A., Kelterborn, S., Evers, H., Kreimer, G., Sizova, I. and Hegemann, P. (2017). Targeting of Photoreceptor Genes in *Chlamydomonas reinhardtii* via Zinc-Finger Nucleases and CRISPR/Cas9. *Plant Cell*. 2498–2518. https://doi.org/10.1105/tpc.17.00659

Grunwald, H. A., Gantz, V. M., Poplawski, G. et al. (2019). Super-Mendelian Inheritance Mediated by CRISPR–Cas9 in the Female Mouse Germline. *Nature*. 566, 105–109. https://doi.org/10.1038/s41586-019-0875-2

Guha, T. K. and Edgell, R. G. (2017). Applications of Alternative Nucleases in the Age of CRISPR/Cas 9. *International Journal of Molecular Sciences*. 18, 2565. https://doi.org/10.3390/ijms18122565

Guha, T. K., Wai, A. and Hausner, G. (2017). Programmable Genome Editing Tools and Their Regulation for Efficient Genome Engineering. *Computational and Structural Biotechnology Journal*. 15, 146–160. https://doi.org/10.1016/j.csbj.2016.12.006.

Guo, T., Feng, Y. L., Xiao, J. J., Liu, Q., Sun, X. N. and Xiang, J. F. (2018). Harnessing Accurate Non-Homologous End Joining for Efficient Precise Deletion in CRISPR/Cas9-Mediated Genome Editing. *Genome Biology*. 19(1), 170. https://doi.org/10.1186/s13059-018-1518-x

Gurdon, J. B. and Melton, D. A. (2008). Nuclear Reprogramming in Cells. *Science*. 322, 1811–1815.

Haeussler, M., Sch€onig, K., Eckert, H., Eschstruth, A., Mianne, J., Renaud, J. B., Schneider-Maunoury, S., Shkumatava, A., Teboul, L., Kent, J., Joly, J. S. and Concordet, J. P. (2016). Evaluation of Off-Target and On-Target Scoring Algorithms and Integration into the Guide RNA Selectiontool CRISPOR. *Genome Biology*. 17, 148.

Hajiahmadi, Z., Movahedi, A., Wei, H., Li, D., Orooji, Y., Ruan, H. and Zhuge, Q. (2019). Strategies to Increase On-Target and Reduce Off-Target Effects of the CRISPR/Cas9 System in Plants. *International Journal of Molecular Sciences*. 20(15), 1–19. https://doi.org/10.3390/ijms20153719

Hammond, A. M. and Galizi, R. (2017). Gene Drives to Fight Malaria: Current State and Future Directions. *Pathogens and Global Health*. 111(8), 412–423, https://doi.org/10.1080/20477724.2018.1438880

Hammond, A. M. and Galizi, R. (2018). Gene Drives to Fight Malaria: Current State and Future Directions. *Pathogens and Global Health*. 111(8), 1–12. https://doi.org/10.1080/20477724.2018.1438880

Hammond, A. M., Galizi, R., Kyrou, K., Simoni, A., Siniscalchi, C., Katsanos, D., Gribble, M., Baker, D., Marois, E., Russell, S., Burt, A., Windbichler, N., Crisanti, A. and Nolan, T. (2016). A CRISPR-Cas9 Gene Drive System Targeting Female Reproduction in the Malaria Mosquito Vector *Anopheles gambiae*. *Nature Biotechnology*. 34(1), 78–83. https://doi.org/10.1038/nbt.3439

Hammond, A. M., Kyrou, K., Bruttini, M., North, A., Galizi, R., Karlsson, X., Kranjc, N., Carpi, F. M., D'Aurizio, R., Crisanti, A. and Nolan, T. (2017). The Creation and Selection of Mutations Resistant to a Gene Drive over Multiple Generations in the Malaria Mosquito. *PLoS Genetics*. 13(10), p.e1007039.

Hammond, A. M., Kyrou, K., Gribble, M., Karlsson, X., Morianou, I., Galizi, R., Beaghton, A., Crisanti, A. and Nolan, T. (2018). Improved CRISPR-Based Suppression Gene Drives Mitigate Resistance and Impose a Large Reproductive Load on Laboratory-Contained Mosquito Populations. *Preprint*. https://doi.org/10.1101/360339

Han, W. and She, Q. (2017). CRISPR History: Discovery, Characterization, and Prosperity. In *Progress in Molecular Biology and Translational Science*. Amsterdam: Elsevier Inc., volume 152. https://doi.org/10.1016/bs.pmbts.2017.10.001

Händel, E. M., Alwin, S. and Cathomen, T. (2009). Expanding or Restricting the Target Site Repertoire of Zinc-Finger Nucleases: The Inter-Domain Linker as a Major Determinant Of Target Site Selectivity. *Molecular Therapy: Journal of the American Society Of Gene Therapy*. 17(1), 104–111. https://doi.org/10.1038/mt.2008.233

Hao, Y. (2014). Genome Editing with Cas9 in Adult Mice Corrects a Disease Mutation and Phenotype. *Nature Biotechnology*. 551–553.

Harumoto, T., Fukatsu, T. and Lemaitre, B. (2018). Common and Unique Strategies of Male Killing Evolved in Two Distinct Drosophila Symbionts. *Proceedings of the Royal Society B: Biological Sciences*. 285(1875), 20172167.

Harvey-Samuel, T., Campbell, K., Edgington, M. and Alphey, L. (2019). *Trialling Gene Drives to Control Invasive Species: What, Where and How?* Veitch, C. R., Clout, M. N., Martin, J. R., Russell, C., West, C. J., Eds. Occasional Paper SSC no. 62. Gland: IUCN, 618–627.

Hatada, I. (2017). Genome Editing in Animals: Methods and Protocols. *Methods in Molecular Biology*. 1630. https://doi.org/10.1007/978-1-4939-7128-2

Haurwitz, R. E., Jinek, M., Wiedenheft, B., Zhou, K. and Doudna, J. A. (2010). Sequence- and Structure-Specific RNA Processing by a CRISPR Endonuclease. *Science*. 329, 1355–1358. [PubMed: 20829488]

Heitman, E., Sawyer, K. and Collins, J. P. (2016). Gene Drives on the Horizon: Issues for Biosafety. *Applied Biosafety*. 21(4), 173–176. https://doi.org/10.1177/1535676016672631

Herrmann, B. G. and Bauer, H. (2012). The Mouse t-Haplotype: A Selfish Chromosome-Genetics Molecular Mechanism and Evollution. In M. Macholán (Ed.), *Evolution of the House Mouse*. Cambridge: Cambridge University Press, 1st edition, 297–314.

Heu, C. C., Luan, J., McCullough, F. M. and Rasgon, J. L. (2020). CRISPR/Cas9-Based Genome Editing in the Silverleaf Whitefly (*Bemisia tabaci*). *bioRxiv*. 3(2), 2020.03.18.996801. https://doi.org/10.1101/2020.03.18.996801

Hille, F. and Charpentier, E. (2016). CRISPR-Cas: Biology, Mechanism, Relevance. *Philosophical Transactions of the Royal Society B: Biological Sciences*. 371(1707), 20150496. https://doi.org/10.1098/rstb.2015.0496

Hoban, M. D. and Bauer, D. E. (2016). A Genome Editing Primer for the Hematologist. *Blood*. 127(21), 2525–2535. https://doi.org/10.1182/blood-2016-01-678151

Hoffman, A. A., Ankery, R., Edward, S., Frammer, M., Hayes, K., Higgins, T. J., Mayo, O., Meek, S., Robin, C., Sheppard, A. and Small, I. (2017). *Implications of Emerging Technologies*. Canberra: Australian Academy of Science.

Horvath, P. and Barrangou, R. (2010). CRISPR/Cas, the Immune System of Bacteria and Archaea. *Science*. 167–170.

Horvath, P. and Barrangou, R. (2012). CRISPR: New Horizons in Phage Resistance and Strain Identification. *Annual Review of Food Science and Technology*. 3, 143–162.

Houben, A. (2017). B Chromosomes—A Matter of Chromosome Drive. *Frontiers in Plant Science*. 8, 210. https://doi.org/10.3389/fpls.2017.00210

Hsu, P. D., Lander, E. S. and Zhang, F. (2014). Development and Applications of CRISPR-Cas9 for Genome Engineering. *Cell*. 157, 1262–1278.

Hu, X. F., Zhang, B., Liao, C. H. and Zeng, Z. J. (2019). High-Efficiency CRISPR/Cas9-mediated Gene Editing in Honeybee (*Apis mellifera*) Embryos. *G3: Genes, Genomes, Genetics*. 9, 1759–1766. https://doi.org/10.1534/g3.119.40013 02

Hu, Y., Xi, Z., Liu, X., Wang, J., Guo, Y., Ren, D., Liu, Q. et al. (2020). Identification and Molecular Characterization of *Wolbachia* Strains in Natural Populations of *Aedes albopictus* in China. *Parasites and Vectors*. 13(1), 28. https://doi.org/10.1186/s13071-020-3899-4

Huang, N., Huang, Z., Gao, M., Luo, Z., Zhou, F., Liu, L. and Xiao, Q. (2018). Induction of Apoptosis in Imatinib Sensitive and Resistant Chronic Myeloid Leukemia Cells by Efficient Disruption of bcr-abl Oncogene with Zinc Finger Nucleases. *Journal of Experimental & Clinical Cancer Research*. 37, 62. https://doi.org/10.1186/s13046-018-0732-4

Ifuku, M., Iwabuchi, K. A., Tanaka, M., Lung, M. S. Y. and Hotta, A. (2018). Restoration of Dystrophin Protein Expression by Exon Skipping Utilizing CRISPR-Cas9 Inmyoblasts Derived from DMD Patient iPS Cells. *Methods in Molecular Biology*. 1828, 191–217.

Ishino, Y., Krupovic, M. and Forterre, P. (2018). History of CRISPR-Cas from Encounter with a Mysterious. *Journal of Bacteriology*. 200(7), e00580.

Jansen, R., Embden, J. D., Gaastra, W. and Schouls, L. M. (2002). Identification of Genes That Are Associated with DNA Repeats in Prokaryotes. *Molecular Microbiology*. 43, 1565–1575.

Jeffries, C. L., Tantely, L. M., Raharimalala, F. N., Hurn, E., Boyer, S. and Walker, T. (2018). Diverse Novel Resident *Wolbachia* Strains in Culicine Mosquitoes from Madagascar. *Scientific Reports*. 8(1), 1–15. https://doi.org/10.1038/s41598-018-35658-z

Ji, H., Lu, P., Liu, B., Qu, X., Wang, Y., Jiang, Z. and Zhu, H. (2018). Zinc-Finger Nucleases Induced by HIV-1 Tat Excise HIV-1 from the Host Genome in Infected and Latently Infected Cells. *Molecular Therapy: Nucleic Acid*. 67–74. https://doi.org/10.1016/j.omtn.2018.04.014

Jiang, F. and Doudna, J. A. (2017). CRISPR—Cas9 Structures and Mechanisms. *Annual Review of Biophysics*. 46(1), 505–529. https://doi.org/10.1146/annurev-biophys-062215-010822

Jiggins, F. M. (2017). The Spread of *Wolbachia* through Mosquito Populations. *PLoS Biology*. 15(6).

Jinek, M., Chylinski, K., Fonfara, I., Hauer, M., Doudna, J. A. and Charpentier, E. (2012). A Programmable Dual-RNA-Guided DNA Endonuclease in Adaptive Bacterial Immunity. *Science*. 337(6096), 816–821. https://doi.org/10.1126/science.1225829

John, M. K. (2020). Potential for a CRISPR Gene Drive to Eradicate or Suppress Globally Invasive Social Wasps. *Genes*. https://doi.org/10.1038/s41598-020-69259-6

John, R., Christian, M. S. and Berckert, V. B. (2013). Origins and Applications of CRISPR-Mediated Genome Editing. *Einstein Journal of Biology and Medicine: EJBM*. 31(1–2), 2–5. https://doi.org/10.23861/ejbm201631754

Johnson, C. (2018). *Briefing for CBD Delegates: Synthetic Gene Drives—Genetic Engineering Gone Wild*. www.synbiogovernance.org/wp-content/uploads/2018/06/5.-BICSBAG_Gene-Drives-Briefing.pdf

Jones, M. S. (2019). *Plants, People and Planet*. https://nph.onlinelibrary.wiley.com/doi/full/10.1002/ppp3.16

Jones, M. S., Delborne, J. A., Elsensohn, J., Mitchell, P. D. and Brown, Z. S. (2019). Does the U.S. Public Support Using Gene Drives in Agriculture? And What Do They Want to Know? *Science Advances*. 5(9). https://doi.org/10.1126/sciadv.aau8462

Joung, J. K. and Sander, J. D. (2013). CRISPR-Cas Systems for Genome Editing, Regulation and Targeting. *NIH Public Access*. 14(1), 49–55. https://doi.org/10.1038/nrm3486. TALENs

Kageyama, D., Ohno, M., Sasaki, T., Yoshido, A., Konagaya, T., Jouraku, A., Kuwazaki, S., Kanamori, H., Katayose, Y., Narita, S. and Miyata, M. (2017). Feminizing *Wolbachia* Endosymbiont Disrupts Maternal Sex Chromosome Inheritance in a Butterfly Species. *Evolution Letters*. 1(5), 232–244.

Kaiser, J. (2016). First Proposed Human Test of CRISPR Passes Initial Safety Review. *Science*. https://www.sciencemag.org/news/2016/06/first-proposed-human-test-crispr-passes-initial-safety-review

Kajtoch, Ł. and Kotásková, N. (2018). Current State of Knowledge on *Wolbachia* Infection among Coleoptera: A Systematic Review. *PeerJ*. 3, 1–31. https://doi.org/10.7717/peerj.4471

Kanchiswamy, C. N. (2016). DNA-Free Genome Editing Methods for Targeted Crop Improvement. *Plant Cell Reports*. 35 1469–1474.10.1007/s00299-016-1982-2

Keller, A. N. (2017). Drugs and Drug-Like Molecules Can Modulate the Function of Mucosal-Associated Invariant T Cells. *Nature Immunology*. 18, 402–411.

Khan, F. A., Pandupuspitasari, N. S., ChunJie, H., Ahmad, H. I., Wang, K., Ahmad, M. J. and Zhang, S. (2018). Applications of CRISPR/Cas9 in Reproductive Biology. *Current Issues in Molecular Biology*. 94–102. https://doi.org/10.21775/cimb.026.093

Khan, S. H. (2019). Genome-Editing Technologies : Concept, Pros, and Cons of Various Genome-Editing Techniques and Bioethical Concerns for Clinical Application. *Molecular Therapy: Nucleic Acid*. 326–334. https://doi.org/10.1016/j.omtn.2019.02.027

Kim, M. S. and Kini, A. G. (2017). Engineering and Application of Zinc Finger Proteins and TALEs for Biomedical Research. *Molecular Cells*. 40(8), 533–541. https://doi.org/10.14348/molcells.2017.0139

Kira, S., Makarova, Y., Wolf, I. and Koonin, V. E. (2019). Classification and Nomenclature of CRISPR—Cas Systems; Where from Here. *CRISPR Journal*. 1, 5.

Kittayapong, P., Kaeothaisong, N. O., Ninphanomchai, S. and Limohpasmanee, W. (2018). Combined Sterile Insect Technique and Incompatible Insect Technique: Sex Separation and Quality of Sterile Aedes Aegypti Male Mosquitoes Released in a Pilot Population Suppression Trial in Thailand. *Parasites and Vectors*. 11(Suppl 2). https://doi.org/10.1186/s13071-018-3214-9

Koch, L. (2016). Genetic Engineering: A New Player in Genome Editing. *Nature Reviews Genetics*. 17(7), 375.

Kodandaramaiah, U. (2008). Effects of *Wolbachia* on Butterfly Life History and Ecology. *Advances*. 16(Ci).

Komor, A. C., Badran, A. H. and Liu, D. R. (2017). CRISPR-Based Technologies for the Manipulation of Eukaryotic Genomes. *Cell*. 168, 20–36. [PubMed]

Koonin, E. V. and Makarova, K. S. (2019). Origins and Evolution of CRISPR-Cas Systems. *Philosophical Transactions of the Royal Society of London: Series B, Biological Sciences*. 374(1772), 20180087. https://doi.org/10.1098/rstb.2018.0087

Kosicki, M., Tomberg, K. and Bradley, A. (2018). Repair of Double-Strand Breaks Induced by CRISPR-Cas9 Leads to Large Deletions and Complex Rearrangements. *Nature Biotechnology*. 36(8), 765–771. https://doi.org/10.1038/nbt.4192

Kruminis-Kaszkiel, E., Juramek, J., Maksymowicz, W. and Wotkiewizc, J. (2018). CRISPR Cas 9 Technology as an Emerging Tool for Targeting Amyotrophic Lateral Sclerosis. *International Journal of Molecular Sciences*. https://doi.org/10.3390/ijms190309

Kunin, V., Sorek, R. and Hugenholtz, P. (2007). Evolutionary Conservation of Sequence and Secondary Structures in CRISPR Repeats. *Genome Biology*. 8, R61. https://doi.org/10.1186/gb-2007-8-4-r61

Kuzma, J., Gould, F., Brown, Z., Collins, J., Delborne, J., Frow, E., Esvelt, K., Guston, D., Leitschuh, D., Oye, K. and Stauffer, S. (2018). A Roadmap for Gene Drives: Using Institutional Analysis and Development to Frame Research Needs and Governance in a Systems Context. *Journal of Responsible Innovation*. 5(Supp 1), S13–S39. https://doi.org/10.1080/23299460.2017.1410344

Kyrou, K., Hammond, A. M., Galizi, R., Kranjc, N., Burt, A., Beaghton, A. K., Nolan, T. and Crisanti, A. (2018). A CRISPR—Cas9 Gene Drive Targeting Doublesex Causes Complete Population Suppression in Caged *Anopheles gambiae* Mosquitoes. *Nature Biotechnology*. 36(11), 1062–1071. https://doi.org/10.1038/nbt.4245

Lander, E. S. (2016). The Heroes of CRISPR. *Cell*. 164.

Lau, W., Fischbach, M. A., Osbourn, A. and Sattely, E. S. (2014). Key Applications of Plant Metabolic Engineering. *PLoS Biology*. 12(6), e1001879. https://doi.org/10.1371/journal.pbio.1001879.

Lee, H. J., Kim, E., Kim, J. S. (2010). Targeted Chromosomal Deletions in Human Cells Using Zinc Finger Nucleases. *Genome Research*. 20, 81–89.

Leftwich, P. T., Edgington, M. P., Harvey-Samuel, T., Carabajal Paladino, L. Z., Norman, V. C. and Alphey, L. (2018). Recent Advances in Threshold-Dependent Gene Drives for Mosquitoes. *Biochemical Society Transactions*. 46(5), 1203–1212. https://doi.org/10.1042/BST20180076

Leggewie, M., Krumkamp, R., Badusche, M., Heitmann, A., Jansen, S., Schmidt-Chanasit, J., . . . Becker, S. C. (2018). *Culex torrentium* Mosquitoes from Germany Are Negative for *Wolbachia*. *Medical and Veterinary Entomology*. 32(1), 115–120. https://doi.org/10.1111/mve.12270

Lei, S., Zhang, F., Yun, Y., Zhou, W. and Peng, Y. (2020). *Wolbachia* Bacteria Affect Rice Striped Stem Borer (*Chilo suppressalis*) Susceptibility to Two Insecticides. *Bulletin of Insectology*. 73(1), 39–44.

Lemay, M. L., Tremblay, D. M. and Moineau, S. (2017). Genome Engineering of Virulent Lactococcal Phages Using CRISPR-Cas9. *ACS Synthetic Biology*. 6(7), 1351–1358.

Lepore, M. (2017). Functionally Diverse Human T Cells Recognize Non-Microbial Antigens Presented by MR1. *eLife*. 6, 1–22.

Li, D., Qiu, Z., Shao, Y. et al. (2013). Heritable Gene Targeting in the Mouse and Rat Using a CRISPR-Cas System. *Nature Biotechnology*. 31, 681–683. https://doi.org/10.1038/nbt.2661

Li, H., Yang, Y., Hong, W., Huang, M., Wu, M. and Zhao, X. (2020). Applications of Genome Editing Technology in the Target Therapy of Human Diseases: Mechanisms, Advances and Prospects. *Sig Transduct Target Therapy*. 5, 1. https//doi.org./10.1038/s41392–019–0089-y

Li, M., Yang, T., Kandul, N. P., Bui, M., Gamez, S., Raban, R., Bennett, J., Lanzaro, G. C., Schmidt, H., Lee, Y. and Marshall, J. M. (2020). Development of a Confinable Gene Drive System in the Human Disease Vector *Aedes aegypti*. *Elife*. 9, e51701.

Liddicoat, J. (2016). *Gene Drive: Regulatory, Legal and Ethical Issues*. Cambridge: Cambridge University Press, 1–14.

Lin, L. and Luo, Y. (2019). Tracking CRISPR's Footprints. *Methods in Molecular Biology*. 1961. https://doi.org/10.1007/978-1-4939-9170-9_2

Lindholm, A. K., Musolf, K., Weidt, A. and Konig, B. (2013). Mate Choice for Genetic Compatibility in the House Mouse. *Ecology and Evolution*. 3(5), 1231–1247. https://doi.org/10.1002/ece3.534.

Lindsey, A. R. I., Rice, D. W., Bordenstein, S. R., Brooks, A. W., Bordenstein, S. R. and Newton, I. L. G. (2018). Evolutionary Genetics of Cytoplasmic Incompatibility Genes cifA and cifB in Prophage WO of *Wolbachia*. *Genome Biology and Evolution*. 10(2), 434–451. https://doi.org/10.1093/gbe/evy012

Loeff, L., Brouns, S. J. J. and Joo, C. 2018. Repetitive DNA Reeling by the Cascade-Cas3 Complex in Nucleotide Unwinding Steps. *Molecular Cell*. 70, 385–394 e383.

Ma, H., Marti-Gutierrez, N., Park, S. W., Wu, J., Lee, Y. and Suzuki, K. (2017). Correction of a Pathogenic Gene Mutation in Human Embryos. *Nature*. 548, 413–419.10.1038/nature23305

Ma, W. J. and Schwander, T. (2017). Patterns and Mechanisms in Instances of Endosymbiont-Induced Parthenogenesis. *Journal of Evolutionary Biology*. 30(5), 868–888.

Macias, V. M., McKeand, S., Chaverra-Rodriguez, D., Hughes, G. L., Fazekas, A., Pujhari, S., Jasinskiene, N., James, A. A. and Rasgon, J. L. (2020). *Cas9-Mediated Gene-Editing in the Malaria Mosquito Anopheles stephensi by ReMOT Control*. https://doi.org/10.1534/g3.120.401133

Macias, V. M., Ohm, J. R. and Rasgon, J. L. (2017). Gene Drive for Mosquito Control: Where Did It Come from and Where Are We Headed? *International Journal of Environmental Research and Public Health*. 14(9). https://doi.org/10.3390/ijerph14091006

Maciel-de-freitas, R., Aguiar, R., Bruno, R. V, Guimarães, M. C., Lourenço-de-oliveira, R., Sorgine, M. H. F., Moreira, L. A. et al. (2012). *Por que se Necesitan Herramientas Alternativas de Control de ETV en America*. Washington, DC: ETV, volume 107, 828–829, September.

Maeder, M. L. and Gersbach, C. A. (2016). Genome-Editing Technologies for Gene and Cell Therapy. Molecular Therapy. *Journal of the American Society of Gene Therapy*. 24(3), 430–446. https://doi.org/10.1038/mt.2016.10

Mahmoudian-sani, M, Fanouish, G., Mahdavinezhad, A. and Saidijam, M. (2017). CRISPR Genome Editing and Its Medical Applications. *Medical Biotechnology*. 280–282.

Mahmoudian-sani, M., Farnoosh, G., Mahdavinezhad, A. and Saidijam, M. (2018). CRISPR Genome Editing and Its Medical Applications. *Biotechnology and Biotechnological Equipment*, 32(2), 286–292.

Makarova, K. S., Grishin, N. V., Shabalina, S. A., Wolf, Y. I. and Koonin, E. V. (2006). A Putative RNA-Interference-Based Immune System in Prokaryotes: Computational Analysis of the Predicted Enzymatic Machinery, Functional Analogies with Eukaryotic RNAi, and Hypothetical Mechanisms Of Action. *Biology Direct*. 1, 7.

Makarova, K. S., Haft, D. H., Barrangou, R., Brouns, S. J., Charpentier, E., Horvath, P., Moineau, S., Mojica, F. J., Wolf, Y. I., Yakunin, A. F., et al. (2011). Evolution and Classification of the CRISPR-Cas Systems. *Nature Reviews Microbiology*. 9, 467–477. [PubMed: 21552286]

Makarova, Y., Kira, S. and Eugene, V. (2018). Origins and Evolution of CRISPR Cas Systems. *Philosophical Transactions of the Royal Society*. 374, 20180087

Mali, P., Yang, L., Esvelt, K. M., Aach, J., Guell, M., DiCarlo, J. E., Norville, J. E. and Church, G. M. (2013). RNA-Guided Human Genome Engineering via Cas9. *Science (New York, N.Y.)*. 339(6121), 823–826. https://doi.org/10.1126/science.1232033

Malnoy, M., Viola, R., Jung, M., Koo, O., Kim, S., Kim, J., Velasco, R. and Kanchiswamy, C. N. (2016). DNA-Free Genetically Edited Grapevine and Apple Protoplast Using CRISPR/Cas9 Ribonucleoproteins. *Front Plant Science*. https://doi.org/10.3389/fpls.2016.01904.

Malzahn, A., Lowder, L. and Qi, Y. (2017). Plant Genome Editing with TALEN and CRISPR. *Cell Bioscience*. 7(1), 21.

Manser, A., Lindholm, A. K., Simmons, L. W. and Firman, R. C. (2017). Sperm Competition Suppresses Gene Drive among Experimentally Evolving Populations of House Mice. *Molecular Ecology*. 20, 5784–5792.

Mao, Y. F., Botella, J. R., Liu, Y. G., Zhu, J. K. (2019). Gene Editing in Plants: Progress and Challenges. *National Science Review*. 6, 421–437.

Mao, Y. F., Zhang, H., Xu, N., Zhang, B., Gou, F. and Zhu, J. K. (2013). Application of the CRISPR-Cas System for Efficient Genome Engineering in Plants. *Molecular Plant*. 6(6), 2008–2011. https://doi.org/10.1093/mp/sst121

Marraffini, L. A. (2015). CRISPR-Cas Immunity in Prokaryotes. *Nature*. 526, 55–61. https://doi.org/10.1038/nature15386

Marraffini, L. A. and Sontheimer, E. J. (2008). CRISPR Interference Limits Horizontal Gene Transfer in Staphylococci by Targeting DNA. *Science*. 322(5909), 1843–1845. https://doi.org/10.1126/science.1165771.

Marshall, J. F. (1938). *The British Mosquitoes*. London: Trustees of the British Museum, 253–254.

Marshall, J. M. (2010). The Cartagena Protocol and Genetically Modified Mosquitoes. *Nature Biotechnology*. 28, 896–897.

Marshall, J. M. (2011). The Cartagena Protocol in the Context of Recent Releases of Transgenic and *Wolbachia*-Infected Mosquitoes. *Asia-Pacific Journal of Molecular Biology and Biotechnology*. 19, 93–100. http://jmarshall.berkeley.edu/MarshallAPJMBB2011.pdf

Marshall, J. M. and Akbari, O. S. (2016). Chapter 9—Gene Drive Strategies for Population Replacement. In *Genetic Control of Malaria and Dengue. Academic Press*, 169–200. ISBN 9780128002469. https://doi.org/10.1016/B978-0-12-800246-9.00009-0.

Marshall, J. M., and Akbari, O. S. (2018). Can CRISPR-Based Gene Drive Be Confined in the Wild? A Question for Molecular and Population Biology. *ACS Chemical Biology*. 16: 13(2), 424–430. https://doi.org/10.1021/acschembio.7b00923. Epub 2018 Feb 7. PMID: 29370514.

Marshall, J. M., Buchman, A. and Akbari, O. S. (2017). Overcoming Evolved Resistance to Population-Suppressing Homing-Based Gene Drives. *Scientific Reports*. 7(1), 3776.

Marshall, J. M. and Hay, B. A. (2011). Inverse *Medea* as a Novel Gene Drive System for Local Population Replacement: A Theoretical Analysis. *Journal of Heredity*. 102(3), 336–341. https://doi.org/10.1093/jhered/esr019

Marshall, J. M. and Hay, B. A. (2014). *Medusa*: A Novel Gene Drive System for Confined Suppression of Mosquito Populations. *PLoS ONE*. 9, e102694. http://journals.plos.org/plosone/article?id=10.1371/journal.pone.0102694

Marshall, J. M., Pittman, G. W., Buchman, A. B. and Hay, B. A. (2011). Semele: A Killer-Male, Rescue-Female System for Suppression and Replacement of Insect Disease Vector Populations. *Genetics*. 187(2), 535–551.

Martínez-Fortún, J., Phillips, D. W. and Jones, H. D. (2017). Potential Impact of Genome Editing in World Agriculture. *Emerg Top Life Science*. 1, 117–133.

Martinez-Lage, M., Puig-Serra, P., Menendez, P., Torres-Ruiz, R. and Rodriguez-Perales, S. (2018). CRISPR/Cas9 for Cancer Therapy: Hopes and Challenges. *Biomedicines*. 12; 6(4), 105. https://doi.org/10.3390/biomedicines6040105. PMID: 30424477; PMCID: PMC6315587

Mashatola, T., Ndo, C., Koekemoer, L. L., Dandalo, L. C., Wood, O. R., Malakoane, L., Munhenga, G. et al. (2018). A Review on the Progress of Sex-Separation Techniques for Sterile Insect Technique Applications against *Anopheles arabiensis*. *Parasites and Vectors*. 11(Suppl 2). https://doi.org/10.1186/s13071-018-3219-4

McClintock, B. (1956). Controlling Elements and the Gene. *Cold Spring Harbor Symposia on Quantitative Biology*. 21. 197–216. ISSN 0091-7451 (Print) 0091-7451 (Linking).

McFarlane, G. R., Whitelaw, C. B. A. and Lillico, S. G. (2018). CRISPR-Based Gene Drives for Pest Control. *Trends Biotechnology*. 36, 130–133. https://doi.org/10.1016/j.tibtech.2017.10.001.

McLaughlin, R. N., Malik, H. S., Levine, J. D., Kronauer, D. J. C. and Dickinson, M. H. (2017). Genetic Conflicts: The Usual Suspects and Beyond. *Journal of Experimental Biology*. 220(1), 6–17. https://doi.org/ https://doi.org/10.1242/jeb.148148

Medina, R. F. (2018). *Gene Drives and the Management of Agricultural Pests*. https://www.tandfonline.com/doi/full/10.1080/23299460.2017.1407913

Mei, T., Liu, C. J., Yang, J., Tai, L. and Zhao, L. (2016). Modified Cas (CRISPR-Associated Proteins) for Genome Editing and Beyond. *Advanced Techniques in Biology & Medicine*. 4, 187. https://doi.org/10.4172/2379-1764.1000187

Meister, G. A. and Grigliatti, T. A. (1993). Rapid Spread of a P Element/Adh Construct Through Experimental Populations of Drosophila Melanogaster. *Genome*. 36, 1169–1175.

Mercier, O. R., KingHunt, A. and Lester, P. J. (2019). Novel Biotechnologies for Eradicating Wasps: Seeking Māori Studies Students' Perspectives with Q Method. *Kōtuitui: NZJ Social Sciences Online*. 14, 136–156. https://doi.org/10.1080/11770 83x.2019.15782 45

Miki, D., Wang, R., Li, J., Kong, D., Zhang, L. and Zhu, J. K. (2021). Gene Targeting Facilitated by Engineered Sequence-Specific Nucleases and Its Potential for Applications in Crop Improvement. *Plant and Cell Physiology*, pcab034. https://doi.org/10.1093/pcp/pcab034

Min, J., Smidler, A. L., Najjar, D. and Esvelt, K. M. (2018). Harnessing Gene Drive. *Journal of Responsible Innovation*. 5(1), S40–S65. https://doi.org/10.1080/23299460.2017.1415586

Mir, A., Edraki, A., Lee, J. and Sontheimer, E. J. (2018). *Type II-C CRISPR-Cas9*. https://pubmed.ncbi.nlm.nih.gov/29202216/

Mishra, N., Shrivastava, N. K. and Nayak, A. (2018). *Wolbachia*: A Prospective Solution to Mosquito Borne Diseases. *International Journal of Mosquito Research*. 5(2), 01–08.

Mohanty, S., Dash, A. and Pradhan, C. K. (2019). CRISPR-Cas9 Technology: A Magical Tool for DNA Editing. *International Journal of Biosciences and Bioengineering*. 0051:1:005.

Mojica, F. J., Diez-Villasenor, C., Garcia-Martinez, J. and Almendros, C. (2009). *Short Motif Sequences Determine the Targets of the Prokaryotic CRISPR Defence System*. https://pubmed.ncbi.nlm.nih.gov/19246744/

Molina, R., Montoya, G. and Prieto, J. (2011). Meganucleases and Their Biomedical Applications. *ELS*. https://doi.org/10.1002/9780470015902.a0023179

Moon, S. B., Kim, D. Y., Ko, J. H. and Kim, Y. S. (2019). Recent Advances in the CRISPR Genome Editing Tool Set. *Experimental and Molecular Medicine*. 51, 130. https://doi.org/10.1038/s12276-019-0339-7

Moore, F. E., Reyon, D., Sander, J. D., Martinez, S. A., Blackburn, J. S., Khayter, C., Joung, J. K. et al. (2012). Improved Somatic Mutagenesis in Zebrafish Using Transcription Activator-Like Effector Nucleases (TALENs). *Cells*. 7(5), 1–9. https://doi.org/10.1371/journal.pone.0037877

Moretti, R., Marzo, G. A., Lampazzi, E. and Calvitti, M. (2018a). Cytoplasmic Incompatibility Management to Support Incompatible Insect Technique against *Aedes albopictus*. *Parasites and Vectors*. 11(Suppl 2). https://doi.org/10.1186/s13071-018-3208-7

Moretti, R., Yen, P. S., Houé, V., Lampazzi, E., Desiderio, A., Failloux, A. B. and Calvitti, M. (2018b). Combining *Wolbachia*-Induced Sterility and Virus Protection to Fight *Aedes albopictus*-Borne Viruses. *PLOS Neglected Tropical Diseases*. 18: 12(7), e0006626. https://doi.org/10.1371/journal.pntd.0006626. PMID: 30020933; PMCID: PMC6066253.

Morhanty, S., Dash, A. and Pradhan, K. C. (2019). *CRISPR Cas 9 Technology: A Magical Tool for DNA Editing*. https://www.researchgate.net/publication/333601605_CRISPR-Cas9_Technology_A_magical_tool_for_DNA_editing

Morisaka, H., Yoshimi, K., Okuzaki, Y., Gee, P., Kunihiro, Y., Sonpho, E., Xu, H., Sasakawa, N., Naito, Y., Nakada, S., Yamamoto, T., Sano, S., Hotta, A., Takeda, J. and Mashimo, T. (2019). CRISPR-Cas3 Induces Broad and Unidirectional Genome Editing in Human Cells. *Nature Communications*. 10(1), 5302. https://doi.org/10.1038/s41467-019-13226-x. PMID: 31811138; PMCID: PMC6897959.

Moyes, C. L., Wiebe, A., Gleave, K., Trett, A., Hancock, P. A., Padonou, G. G., Coleman, M. et al. (2019). Analysis-Ready Datasets for Insecticide Resistance Phenotype and Genotype Frequency in African Malaria Vectors. *Scientific Data*. 6(1), 121. https://doi.org/10.1038/s41597-019-0134-2

Mudziwapasi, R., Nyamusamba, R. P., Mutengwa, T. T., Maphosa, M., Jomane, F. N. and Ndudzo, A. (2018). Unlocking the Potential of CRISPR Technology for Improving Livelihoods in Africa. *Biotechnology and Genetic Engineering Reviews*. 34(2), 198–215. https://doi.org/10.1080/02648725.2018.1482101

Muller, H. J. (1927). Artificial Transmutation of the Gene. *Science*. 66, 84–87.

Murugan, K., Babu, K., Sundaresan, R., Rajan, R. and Sashital, D. G. (2017). The Revolution Continues: Newly Discovered Systems Expand the CRISPR-Cas Toolkit. *Molecular Cell*. 68(1), 15–25. https://doi.org/10.1016/j.molcel.2017.09.007

Nakade, S., Yamamoto, T. and Sakuma, T. (2017). Cas9, Cpf1 and C2c1/2/3—What's next? *Bioengineered*. 8, 265–273.

Nash, A., Urdaneta, G. M., Beaghton, A. K., Hoermann, A., Papathanos, P. A., Christophides, G. K. and Windbichler, N. (2019). Integral Gene Drives for Population Replacement. *Biology Open*. 8(1), 1–11. https://doi.org/10.1242/bio.037762

Nelson, C. E., Hakim, C. H. and Ousterout, D. G. (2016). In Vivo Genome Editing Improves Muscle Function in a Mouse Model of Duchenne Muscular Dystrophy. *Science*. 351, 403–407.

Nestor, M. W. and Wilson, R. L. (2020). Beyond Mendelian Genetics: Anticipatory Biomedical Ethics and Policy Implications for the Use of CRISPR Together with Gene Drive in Humans. *Journal of Bioethical Inquiry*. https://doi.org/10.1007/s11673-019-09957-7

Neve, P. (2018). Gene Drive Systems: Do They Have a Place in Agricultural Weed Management? *Pest Management Science*. 74(12), 2671–2679. https://doi.org/10.1002/ps.5137

Newton, I. L. G. and Slatko, E. (2019). Symbiosis Comes of Age at the 10th Biennial Meeting of *Wolbachia* Researchers. *Applied and Environmental Microbiology*. 85(8), 1–9.

Noble, C., Adlam, B. and Church, G. M. (2017). *Current CRISPR Gene Drive Systems Are Likely to Be Highly Invasive in Wild Populations*, 9–12. https://elifesciences.org/articles/33423

Noble, C., Olejarz, J., Esvelt, K. M., Church, G. M. and Nowak, M. A. (2017). Evolutionary Dynamics of CRISPR Gene Drives. *Science Advances*. 3(4), p.e1601964.

Noman, A., Aqeel, M. and He, S. (2016). CRISPR-Cas9: Tool for Qualitative and Quantitative Plant Genome Editing. *Front Plant Science*. https://doi.org/10.3389/fpls.2016.01740.

North, A. R., Burt, A. and Godfray, H. C. J. (2019). Modelling the Potential of Genetic Control of Malaria Mosquitoes at National Scale. *BMC Biology*. 17(1), 26.

Norwegian Biotechnology Advisory Board. (2017). *Statement on Gene Drives*, 1–18. www.bioteknologiradet.no/filarkiv/2017/02/Statement-on-gene-drives.pdf

Oberhofer, G., Ivy, T. and Hay, B. A. (2018). Behavior of Homing Endonuclease Gene Drives Targeting Genes Required for Viability or Female Fertility with Multiplexed Guide RNAs. *Proceedings of the National Academy of Sciences*. 115 (40) E9343-E9352; https://doi.org/10.1073/pnas.1805278115

Odongo, J. (2019). Genome Editing and Frontiers in Bio-Engineering. *International Journal of Research and Innovation in Applied Science*. IV(IX). ISSN 2454-6194

Ogura, T. and Hiraga, S. (1983). Mini-F Plasmid Genes That Couple Host Cell Division to Plasmid Proliferation. *Proceedings of the National Academy of Sciences of the United States of America*. 80, 4784–8.

Oishi, I., Yoshii, K., Miyahara, D., Kagami, H. and Tagami, T. (2016). Targeted Mutagenesis in Chicken Using CRISPR/Cas9 System. *Scientific Reports*. 6, 23980.

O'Neill, S. L., Ryan, P. A., Turley, A. P., Wilson, G., Retzki, K., Iturbe-Ormaetxe, I., . . . Simmons, C. P. (2018). Scaled Deployment of *Wolbachia* to Protect the Community from Dengue and Other *Aedes* Transmitted Arboviruses. *Gates Open Research*. 2, 36. https://doi.org/10.12688/gatesopenres.12844.2

Osborn, M. J., Webber, B. R., Knipping, F., Lonetree, C. L., Tennis, N., DeFeo, A. P., McElroy, A. N., Starker, C. G., Lee, C., Merkel, S., Lund, T. C., Kelly-Spratt, K. S., Jensen, M. C., Voytas, D. F., Von Kalle, C., Schmidt, M., Gabriel, R., Hippen, K. L., Miller, J. S., . . . Blazar, B. R. (2016). Evaluation of TCR Gene Editing Achieved by TALENs, CRISPR/Cas9, and MegaTAL Nucleases. *Molecular Therapy*. 24(3), 570–581. https://doi.org/10.1038/mt.2015.197

Pan, C., Ye, L., Qin, L., Liu, X., He, Y., Wang, J., Chen, L. and Lu, G. (2016). CRISPR/Cas9-Mediated Efficient and Heritable Targeted Mutagenesis in Tomato Plants in the First and Later Generations. *Scientific Reports*. 6, 24765. https://doi.org/10.1038/srep24765.

Pan, X., Pike, A., Joshi, D., Bian, G., McFadden, M. J., Lu, P., Xi, Z. et al. (2018). The Bacterium *Wolbachia* Exploits Host Innate Immunity to Establish a Symbiotic Relationship with the Dengue Vector Mosquito *Aedes aegypti*. *ISME Journal*. 12(1), 277–288. https://doi.org/10.1038/ismej.2017.174

Parmentier, H. K., Cahaner, A., Poel, J. J. van der, Lamont, S. J. and Pinard-van der Laan, M. H. (2009). Selection for Disease Resistance: Direct Selection on the Immune Response. *Poultry Genetics, Breeding and Biotechnology*. https://doi.org/10.1079/9780851996608.0399

Paschon, D. E., Lussier, S., Wangzor, T., Xia, D. F., Li, P. W., Hinkley, S. J., Rebar, E. J. (2019). Diversifying the Structure of Zinc Finger Nucleases for High-Precision Genome Editing. *Nature Communication*. 10, 1133. https://doi.org/10.1038/s41467-019-08867-x

Pavletich, N. P. and Pabo, C. O. (1991). Zinc Finger-DNA Recognition: Crystal Structure of a Zif268-DNA Complex at 2.1 a Resolution. *Science*. 252, 809–817.

Payne, P., Geyrhofer, L., Barton, N. H. and Bollback, J. P. (2018). CRISPR-Based Herd Immunity Can Limit Phage Epidemics in Bacterial Populations. *eLife*. 7, e32035. https://doi.org/10.7554/eLife.32035.

Pearson, B. M., Louwen, R., van Baarlen, P. and van Vliet, A. H. M. (2015). *Conservation of σ28-Dependent Non-Coding RNA Paralogs and Predicted σ54-Dependent Targets in Thermophilic Campylobacter Species*. https://journals.plos.org/plosone/article?id=10.1371/journal.pone.0141627

Peng, H., Zheng, Y., Blumenstein, M., Tao, D. and Li, J. (2018). *CRISPR Cas 9 Cleavage Efficiency Regression Through Boosting Algorithms and Marker Sequence Profiling*. Sydney, Australia: Faculty of Engineering and Information Technology, University of Technology.

Pennisi, E. (2013). The CRISPR Craze. *Science*. 341, 833–836.

Perlmutter, J. I., Bordenstein, S. R., Unckless, R. L., LePage, D. P., Metcalf, J. A., Hill, T., . . . Bordenstein, S. R. (2019). The Phage Gene WMK Is a Candidate for Male Killing by a Bacterial Endosymbiont. *PLoS Pathogens*. 15(9), 1–29. https://doi.org/10.1371/journal.ppat.1007936

Petersen, B. (2017). Basics of Genome Editing Technology and Its Application in Livestock Species. *Reprod Domest Anim*. 52(Suppl 3), 4–13. https://doi.org/10.1111/rda.13012. PMID:28815851.

Ping, G. (2017). *Invasive Species Management on Military Lands: Clustered Regularly Interspaced Short Palindromic Repeat/CRISPR-Associated Protein 9 (CRISPR/Cas9)-Based Gene Drives*. Vicksburg: US Army Engineer Research and Development Center (ERDC) Environmental Laboratory.

Porro, F., Bockor, L., De Caneva, A., Bortolussi, G. and Muro, A. F. (2014). Generation of Ugt1-Deficient Murine Liver Cell Lines Using TALEN Technology. *PloS One*. 9(8), e104816. https://doi.org/10.1371/journal.pone.0104816.

Porteus, M. (2016). Genome Editing. *A New Approach to Human Therapeutics*. 56, 156–190.

Porteus, M. (2019). A New Class of Medicine Through DNA Editing. *The New England Journal of Medicine*. 380, 947–959.

Pourcel, C. G., Salvignol, G. and Vergnaud, G. (2005). CRISPR Elements in Yersinia Pestis Acquire New Repeats by Preferential Uptake of Bacteriophage DNA, and Provide Additional Tools for Evolutionary Studies. *Microbiology*. 151, 653–663.

Pratap, D. and Sharma, P. (2016). Plant Genome Editing Transcription Activator-Like Effector Nucleases (TALENs): An Efficient Tool for Plant Genome Editing. *Engineering in Life Sciences*. 16(4) https://doi.org/10.1002/elsc.201500126

Proudfoot, C., Carlson, D. F., Huddart, R., Long, C. R., Pryor, J. H. and King, T. J. (2015). Genome Edited Sheep and Cattle. *Transgenic Research*. 24(1), 147–153.

Prowse, T. A. A., Adikusuma, F., Cassey, P., Thomas, P. and Ross, J. V. (2019). A Y-Chromosome Shredding Gene Drive for Controlling Pest Vertebrate Populations. *eLife*. 8, e41873. https://doi.org/10.7554/eLife .41873.

Prowse, T. A. A., Cassey, P., Ross, J. V., Pfitzner, C., Wittmann, T. A. and Thomas, P. (2017). Dodging Silver Bullets: Good CRISPR Gene-Drive Design Is Critical for Eradicating Exotic Vertebrates. *Proceedings of the Royal Society B: Biological Sciences*. 284(1860). https://doi.org/10.1098/rspb.2017.0799

Pu, J., Frescas, D., Zhang, B. and Feng, J. (2015). Utilization of TALEN and CRISPR/Cas9 Technologies for Gene Targeting and Modification. *Experimental Biology and Medicine (Maywood, N.J.)*. 240(8), 1065–1070. https://doi.org/10.1177/1535370215584932

Puchta, H. (2005). The Repair of Double-Strand Breaks in Plants: Mechanisms and Consequences for Genome Evolution. *J Exp Bot*. 56, 1–14.

Pul, Ü., Wurm, R., Arslan, Z., Geißen, R., Hofmann, N. and Wagner, R. (2010). Identification and Characterization of *E. Coli* CRISPR-Cas Promoters and Their Silencing by H-NS. *Molecular Microbiology*. 75(6), 1495–1512.

Qi, Y. (2017). Genome Editing Is Revolutionizing Biology. *Cell & Bioscience*. 7, 35.

Raitskin, O., Schudoma, C., West, A. and Patron, N. J. (2019). *Comparison of Efficiency and Specificity of CRISPR-Associated (Cas) Nucleases in Plants: An Expanded Toolkit for Precision Genome Engineering*. https://journals.plos.org/plosone/article?id=10.1371/journal.pone.0211598

Ran, Y., Patron, N., Kay, P., Wong, D., Buchanan, M., Cao, Y. and Hayden, M. J. (2018). *Zinc Finger Nuclease-Mediated Precision Genome Editing of an Endogenous Gene in Hexaploid Bread Wheat (Triticum aestivum) Using a DNA Repair Template*. 2088–2101. https://doi.org/10.1111/pbi.12941

Rasgon, J. L. and Gould, F. (2005). Transposable Element Insertion Location Bias and the Dynamics of Gene Drive in Mosquito Populations. *Insect Molecular Biology*. 14, 493–500.

Rath, D., Amlinger, L., Bindal, G. and Lundgren, M. (2018). *Type I-E CRISPR-Cas System as Immune System in a Eukaryote*. www.biorxiv.org/content/10.1101/357301v1

Rath, D., Amlinger, L., Rath, A. and Landgrem, M. (2015). The CRISPR-Cas Immune System: Biology Mechanisms and Applications. *Biochimie*. 117, 119–128. https://doi.org/10.1016/j.biochi.2015.03.025

Ray, A. and Felice, D. R. (2015). *Molecular Simulations Have Boosted Knowledge of CRISPR/Cas 9: A Review*. Department of Physics and Astronomy. https://www.riverpublishers.com/journal_read_html_article.php?j=JSAME/7/1/3

Razzaq, A. and Masood, A. (2018). CRISPR/Cas9 System: A Breakthrough in Genome Editing. *Molecular Biology*. 7(2), 1–4. https://doi.org/10.4172/2168-9547.1000210

Rémy, S., Tesson, L., Ménoret, S., Usal, C., Scharenberg, A. M. and Anegon, I. (2010). Zinc-Finger Nucleases: A Powerful Tool for Genetic Engineering of Animals. *Transgenic Res*. 19(3), 363–371. https://doi.org/10.1007/s11248-009-9323-7. Epub 2009 Sep 26. PMID:19821047.

Reveillaud, J., Bordenstein, S. R., Cruaud, C., Shaiber, A., Esen, Ö. C., Weill, M., . . . Eren, A. M. (2019). The *Wolbachia* Mobilome in Culex Pipiens Includes a Putative Plasmid. *Nature Communications*. 10(1). https://doi.org/10.1038/s41467-019-08973-w

Reynolds, L. A., Hornett, E. A., Jiggins, C. D. and Hurst, G. D. D. (2019). Suppression of *Wolbachia*-Mediated Male-Killing in the Butterfly Hypolimnas Bolina Involves a Single Genomic Region. *PeerJ*. 10, 1–10. https://doi.org/10.7717/peerj.7677

Robb, G. B. (2019). Genome Editing with CRISPR-Cas: An Overview. *Current Protocols Essential Laboratory Techniques*, 19, e36. https://doi.org/10.1002/cpet.36

Rode, N. O., Estoup, A., Bourguet, D., Courtier, V. and Florence, O. (2019). Population Management Using Gene Drive : Molecular Design, Models of Spread Dynamics and Assessment of Ecological Risks. *Conservation Genetics*. 20(4), 671–690. https://doi.org/10.1007/s10592-019-01165-5

Rodriguez, E. (2017). Ethical Issues in Genome Editing for Non-Human Organisms Using CRISPR/Cas9 System. *Journal Clin Research Bioethics*. 8, 1000300. https://doi.org/10.4172/2155-9627.1000300

Rodríguez-Leal, D., Lemmon, Z. H., Man, J., Bartlett, M. E. and Lippman, Z. B. (2017). Engineering Quantitative Trait Variation for Crop Improvement by Genome Editing. *Cell*. 171(2), 470–480.e8.

Roggenkamp, E., Giersch, R. M., Schrock, M. N., Turnquist, E., Halloran, M. and Finnigan, G. C. (2018). Tuning CRISPR-Cas9 Gene Drives in *Saccharomyces cerevisiae*. *G3: Genes, Genomes, Genetics*. 8(3), 999–1018. https://doi.org/10.1534/g3.117.300557

Romeis, J., Collatz, J., Glandorf, D. C. M. and Bonsall, M. B. (2020). The Value of Existing Regulatory Frameworks for the Environmental Risk Assessment of Agricultural Pest Control Using Gene Drives. *Environmental Science and Policy*. 108, 19–36. https://doi.org/10.1016/j.envsci.2020.02.016

Ross, P. A., Axford, J. K., Richardson, K. M., Endersby-Harshman, N. M. and Hoffmann, A. A. (2017). Maintaining *Aedes aegypti* Mosquitoes Infected with *Wolbachia*. *Journal of Visualized Experiments*. 126, 1–8. https://doi.org/10.3791/56124

Ross, P. A., Ritchie, S. A., Axford, J. K. and Hoffmann, A. A. (2019). Loss of Cytoplasmic Incompatibility in *Wolbachia*-Infected *Aedes aegypti* under Field Conditions. *PLoS Neglected Tropical Diseases*. 13(4), p.e0007357.

The Royal Society. (2016). *GM Plants: Questions and Answers*. London: Royal Society.

The Royal Society. (2018). *Gene Drive Research—Why It Matters*. London: Royal Society. https://doi.org/10.4324/9780203084601

Rozov, S. M., Permyakova, N. V. and Deineko, E. V. (2019). The Problem of the Low Rates of CRISPR/Cas9-Mediated Knock-ins in Plants: Approaches and Solutions. *International Journal of Molecular Sciences*. 20(13), 3371. https://doi.org/10.3390/ijms20133371

Rubin, G. and Spradling, A. (1982). Genetic Transformation of Drosophila with Transposable Element Vectors. *Science*. 218(4570), 348–353.

Runge, J. N. and Lindholm, A. K. (2018). Carrying a Selfish Genetic Element Predicts Increased Migration Propensity in Free-Living Wild House Mice. *Proceedings. Biological Sciences*, *285*(1888), 20181333. https://doi.org/10.1098/rspb.2018.1333

Sánchez, C. H. M., Bennett, J. B., Wu, S. L., Rašić, G., Akbari, O. S. and Marshall, J. M. (2020). Modeling Confinement and Reversibility of Threshold-Dependent Gene Drive Systems in Spatially-Explicit *Aedes aegypti* Populations. *BMC Biology*.18, 1–14.

Sapranauskas, R., Gasiunas, G., Fremaux, C., Barrangou, R., Horvath, P. and Siksnys, V. (2011). The *Streptococcus thermophilus* CRISPR/Cas System Provides Immunity in *Escherichia coli*. *Nucleic Acids Research*. 39(21), 9275–9282. https://doi.org/10.1093/nar/gkr606

Sarkar, S. (2018). Researchers Hit Roadblocks with Gene Drives. *BioScience*. 68, 474–480. https://doi.org/10.1093/biosci/biy060

Sather, B. D., Ibarra, G. S. R., Sommer, K., Curinga, G., Hale, M., Khan, I. F., Singh, S., Song, Y., Gwiazda, K., Sahni, J., Jarjour, J., Astrakhan, A., Wagner, T. A., Scharenberg, A. M. and Rawlings, D. J. (2015). Efficient Modification of CCR5 in Primary Human Hematopoietic Cells Using a megaTAL Nuclease and AAV Donor Template. *Science Translational Medicine*. 7(307). https://doi.org/10.1126/scitranslmed.aac5530

Schaap, S. (2018). *CRISPR and Animals: Implications of Genome Editing for Policy and Society*, 1–105. www.cogem.net/index.cfm/en/publications/publication/crispr-animals-implications-of-genome-editing-for-policy-and-society?

Schaefer, K. A., Wu, W., Colgan, D. F., Tsang, S. H., Bassuk, A. G. and Mahajan, V. B. (2017). Unexpected Mutations after CRISPR—Cas9 Editing in Vivo. *Nature Methods*. 14, 547–548.

Schebeck, M., Feldkirchner, L., Stauffer, C. and Schuler, H. (2019). Dynamics of an Ongoing *Wolbachia* Spread in the European Cherry Fruit Fly, *Rhagoletis cerasi* (Diptera: Tephritidae). *Insects*. 10(6), 1–11. https://doi.org/10.3390/insects10060172

Schinirring, L. (2016). *'Wolbachia' Efforts Ramp Up to Fight Zika in Brazil, Colombia*. www.cidrap.umn.edu/news-perspective/2016/10/*Wolbachia*-efforts-ramp-fight-zika-brazil-colombia [Accessed May 3, 2020].

Sebastiano, V., Maeder, M. L., Angstman, J. F., Haddad, B., Khayter, C., Yeo, D. T., Goodwin, M. J., Hawkins, J. S., Ramirez, C. L., Batista, L. F., Artandi, S. E., Wernig, M. and Joung, J. K. (2011). In Situ Genetic Correction of the Sickle Cell Anemia Mutation in Human Induced Pluripotent Stem Cells Using Engineered Zinc Finger Nucleases. *Stem Cells (Dayton, Ohio)*. 29(11), 1717–1726. https://doi.org/10.1002/stem.718

Selle, K. and Barrangou, R. (2015). CRISPR-Based Technologies and the Future of Food Science. *Journal of Food Science*. 80(11), R2367-72. https://doi.org/10.1111/1750-3841.13094. Epub 2015 Oct 7. PMID: 26444151.

Shabbir, M. A., Hao, H., Shabbar, M. Z., Hassan, I. H., Igbalz, Z., Ahmed, S., Sattar, A., Igbal, M., Li, J. and Yuan, Z. (2016). *Survival and Evolution of CRISPR–Cas System in Prokaryotes and Its Applications*. https://www.ncbi.nlm.nih.gov/pmc/articles/PMC5035730/

Shao, M., Xu, T. R. and Chen, C. S. (2016). The Big Bang Of Genome Editing Technology: Development and Application of the CRISPR/Cas9 System in Disease Animal Models. *Dongwuxue Yanjiu*. 37(4), 191–204. https://doi.org/10.13918/j.issn.2095-8137.2016.4.191. PMID: 27469250; PMCID: PMC4980067.

Shariat, N., DiMarzio, M. J., Yin, S., Dettinger, L., Sandt, C. H., Lute, J. R. . . . Dudley, E. G. (2013). The Combination of CRISPR-MVLST and PFGE Provides Increased Discriminatory Power for Differentiating Human Clinical Isolates of Salmonella Enterica Subsp. Enterica Serovar Enteritidis. *Food Microbiology*. 34, 164–167.

Shin, H. Y., Wang, C., Lee, H. K., Yoo, K. H., Zeng, X. and Kuhns, T. (2017). CRISPR/Cas9 Targeting Events Cause Complex Deletions and Insertions at 17 Sites in the Mouse Genome. *Nature Communication*. https://doi.org/10.1038/ncomms15464

Shropshire, J. D. and Bordenstein, S. R. (2019). Two-by-One Model of Cytoplasmic Incompatibility: Synthetic Recapitulation by Transgenic Expression of Cifa and Cifb in Drosophila. *PLoS Genetics*. 15(6), 1–20. https://doi.org/10.1371/journal.pgen.1008221

Shukla-Jones, A., Friedrichs, S. and Winickoff, D. (2018). *Gene Editing in an International Context*. https://www.oecd-ilibrary.org/industry-and-services/gene-editing-in-an-international-context_38a54acb-en

Sicard, M., Bonneau, M. and Weill, M. (2019). *Wolbachia* Prevalence, Diversity, and Ability to Induce Cytoplasmic Incompatibility in Mosquitoes. *Current Opinion in Insect Science*. 34, 12–20. https://doi.org/10.1016/j.cois.2019.02.005

Simon, S., Otto, M. and Engelhard, M. (2018). Synthetic Gene Drive: Between Continuity and Novelty. *EMBO Reports*. 19(5), 1–4. https://doi.org/10.15252/embr.201845760

Sinkins, S. P. and Gould, F. (2006). Gene Drive Systems for Insect Disease Vectors. *Nature Reviews Genetics*. 1, 427–435. https://doi.org/10. 1038/nrgl870

Smith, R. C. and Atkinson, P. W. (2011). Mobility Properties of the Hermes Transposable Element in Transgenic Lines of *Aedes aegypti*. *Genetica*. 139(1), 7–22. https://doi.org/10.1007/s10709-010-9459-7. Epub 2010 Jul 3. PMID: 20596755; PMCID: PMC3030943.

Soldner, F. (2016). Parkinson-Associated Risk Variant in Distal Enhancer of α-Synuclein Modulates Target Gene Expression. *Nature*. 533(7601), 95–99.

Sorek, R., Lawrence, M. C. and Wiedenheft, B. (2013). CRISPR-Mediated Adaptive Immune Systems in Bacteria and Archaea. *Annual Review of Biochemistry*. Vol. 82, 237–266. https://doi.org/10.1146/annurev-biochem-072911-172315.

Spaak, P. H. (2019). *The Science and Ethics of Gene Drive TECHNOLOGY CASE Study: Eradicating Malaria Participants' Booklet*. www.europarl.europa.eu/stoa/en/events/farming-without-agro-chemicals

Steffann, J., Jouannet, P., Bonnefont, J. P., Chneiweiss, H. and Frydman, N. (2016). Could Failure in Preimplantation Genetic Diagnosis Justify Editing the Human Embryo Genome? *Cell Stem Cell*. 22(4), 481–482. https://doi.org/10.1016/j.stem.2018.01.004

Steinbrecher, R., Wells, M., Brandt, R., Bucking, E. and Gurian-Sherman, D. (2018). *Gene Drives: A Report on Their Science, Applications, Ethics and Regulations Potential Applications and Risks*. https://ensser.org/publications/2019-publications/gene-drives-a-report-on-their-science-applications-social-aspects-ethics-and-regulations/

Stella, S. and Mantoya, G. (2015). The Genome Editing Revolution: A CRISPR- Cas TALE Off Target Story. *Inside the Cell*. 1(1). https://doi.org/10.1002/icl3.1038

Sumner, S., Law, G. and Cini, A. (2018). Why We Love Bees and Hate Wasps. *Ecol Entomology*. 43, 836–845. https://doi.org/10.1111/een.12676.

Swiech, L., Heidenreich, M., Banerjee, A., Habib, N., Li, Y., Trombetta, J., Sur, M. and Zhang, F. (2015). In Vivo Interrogation of Gene Function in the Mammalian Brain Using CRISPR-Cas9. *Nature Biotechnology*. 33(1), 102–6. https://doi.org/10.1038/nbt.3055. Epub 2014 Oct 19. PMID: 25326897; PMCID: PMC4492112.

Tabebordbar, M., Zhu, K. and Cheng, J. K. (2016). In Vivo Gene Editing in Dystrophic Mouse Muscle and Muscle Stem Cells. *Science*. 351, 407–411.

Takeuchi, R., Choi, M. and Stoddard, B. L. (2014). Engineering Meganucleases for Gene Modification. *Proceedings of the National Academy of Sciences*. 111 (11), 4061–4066. https://doi.org/10.1073/pnas.1321030111

Tan, C. H., Wong, P. S. J., Li, M. I., Yang, H. T., Ng, L. C. and O'Neill, S. L. (2017). wMel Limits Zika and Chikungunya Virus Infection in a Singapore *Wolbachia*-Introgressed Ae. Aegypti

Strain, wMel-Sg. *PLoS Neglected Tropical Diseases*. 11(5), 1–10. https://doi.org/10.1371/journal.pntd.0005496

Tang, X., Ren, Q., Yang, L., Bao, Y., Zhong, Z., He, Y. et al. (2019). Single Transcript Unit CRISPR 2.0 Systems for Robust Cas9 and Cas12a Mediated Plant Genome Editing. *Plant Biotechnology Journal*. 17, 1431–1445. https://doi.org/10.1111/pbi.13068

Tantowijoyo, W., Andari, B., Arguni, E., Budiwati, N., Nurhayati, I., Fitriana, I., . . . O'Neill, S. L. (2020). Stable Establishment of wMel *Wolbachia* in *Aedes aegypti* Populations in Yogyakarta, Indonesia. *PLOS Neglected Tropical Diseases*. 14(4), e0008157. https://doi.org/10.1371/journal.pntd.0008157

Target-Malaria. (2020). *Our Work at Target Malaria*. https://targetmalaria.org/our-work/

Taylor, M. J., Bordenstein, S. R. and Slatko, B. (2018). Microbe Profile: *Wolbachia*: A Sex Selector, a Viral Protector and a Target to Treat Filarial Nematodes. *Microbiology (United Kingdom)*. 164(11), 1345–1347. https://doi.org/10.1099/mic.0.000724

Teem, J. L., Ambali, A., Glover, B. Ouedraogo, J., Makinde, D. and Roberts, A. (2019). Problem Formulation for Gene Drive Mosquitoes Designed to Reduce Malaria Transmission in Africa: Results from Four Regional Consultations 2016–2018. *Malar Journal*. 18, 347 (2019). https://doi.org/10.1186/s12936-019-2978-5

Thomas, K. R., Folger, K. R. and Capecchi, M. R. (1986). High Frequency Targeting of Genes to Specific Sites in the Mammalian Genome. *Cell*. 44, 419–428.

Thompson, P. B. (2018). The Roles of Ethics in Gene Drive Research and Governance. *Journal of Responsible Innovation*. 5, S159–S179. https://doi.org/10.1080/23299460.2017.1415587

Tilman, D., Balzer, C., Hill, J., Befort, B. L. (2011). Global Food Demand and the Sustainable Intensification of Agriculture. *Proceedings of the National Academy of Sciences of the United States of America*. 108, 20260–20264.

Tong, Y., Whitford, M. C., Robertsen, L. H., Blin, K., Jorgensen, S. T., Kiltgard, A. K. et al. (2019). *Highly Efficient DSB-Free Base Editing for Streptomycetes with CRISPR-BEST*. https://www.ncbi.nlm.nih.gov/pmc/articles/PMC6789908/

Townson, J. (2017). Recent Developments in Genome Editing for Potential Use in Plants. *Bioscience Horizons: The International Journal of Student Research*. 10. https://doi.org/10.1093/biohorizons/hzx016

Tycko, J., Myer, V. E. and Hsu, P. D. (2016). Methods for Optimizing CRISPR-Cas9 Genome Editing Specificity. *Molecular Cell*. 63(3), 355–370.

Ün, Ç., Schultner, E., Manzano-Marin, A., Florez, L. V., Seifert, B., Huinze, J. and Oettler, J. (2019). Cytoplasmic Incompatibility Between Old and New World Populations of a Tramp Ant. *Statistical Field Theory*. 53(9), 1689–1699. https://doi.org/10.1017/CBO9781107415324.004

Unckless, R. L., Clark, A. G. and Messer, P. W. (2017). Evolution of Resistance Against CRISPR/Cas9 Gene Drive. *Genetics*. 205(2), 827–841. https://doi.org/10.1534/genetics.116.197285

Urnov, F. D. (2018). Ctrl-alt-inDel: Genome Editing to Reprogram a Cell in the Clinic. *Current Opinion in Genetics and Development*. 52, 48–56.

Urnov, F. D., Rebar, E. J., Holmes, M. C., Zhang, H. S. and Gregory, P. D. (2010). Genome Editing with Engineered Zinc Finger Nucleases. *Natures Reviews/Genetics*. 11.

van der Vlugt, C., van den Akker, E., Roesink, C. H. and Westra, J. (2018). *Risk Assessment Method for Activities Involving Organisms with a Gene Drive Under Contained Use*. RIVM Letter Report 2018-0090. www.rivm.nl/bibliotheek/rapporten/2018-0090.pdf

Vlugt, J. B., Van Der, Brown, D. D., Lehmann, K., Leunda, A. and Willemarck, N. (2018). A Framework for the Risk Assessment and Management of Gene Drive Technology. *Contained Use*. 23(1), 25–31. https://doi.org/10.1177/1535676018755117

Vujošević, M., Rajičić, M. and Blagojević, J. (2018). B Chromosomes in Populations of Mammals Revisited. *Genes (Basel)*. 9: 9(10), 487. https://doi.org/10.3390/genes9100487. PMID: 30304868; PMCID: PMC6210394.

Wang, H. Y. and Yang, H. (2019). Gene-Edited Babies: What Went Wrong and What Could Go Wrong. *PLoS Biology*. 17(4), e3000224. https://doi.org/10.1371/journal.pbio.3000224

Wang, M., Glass, Z. and Xu, Q. (2017). Non-Viral Delivery of Genome-Editing Nucleases for Gene Therapy. *Gene Ther*. 24, 144–150. https://doi.org/10.1038/gt.2016.72

Wang, R. L., Stec, A., Hey, J., Lukens, L. and Doebley, J. (1999). The Limits of Selection During Maize Domestication. *Nature*. 398, 236–239.

Wang, Y., Cheng, X., Shan, Q., Zhang, Y., Liu, J., Gao, C. and Qiu, J. (2014). Simultaneous Editing of Three Homoeoalleles in Hexaploid Bread Wheat Confers Heritable Resistance to Powdery Mildew. *Nature Biotechnology*. 32, 947–951. https://doi.org/10.1038/nbt.2969.

Wang, Y., Zhou, X. Y., Xiang, P. Y., Wang, L. L., Tang, H., Xie, F., Li, L. and Wei, H. (2014). The Meganuclease I-SceI Containing Nuclear Localization Signal (NLS-I-SceI) Efficiently Mediated Mammalian Germline Transgenesis via Embryo Cytoplasmic Microinjection. *PloS One*. 9(9), e108347. https://doi.org/10.1371/journal.pone.0108347

Ward, C. M., Su, J. T., Huang, Y., Lloyd, A. L., Gould, F. and Hay, B. A. (2010). Medea Selfish Genetic Elements as Tools for Altering Traits of Wild Populations: A Theoretical Analysis. *Evolution*. https://doi.org/10.1111/j.1558-5646.2010.01186.x

Ward, C. M., Su, J. T., Huang, Y., Lloyd, A. L., Gould, F. and Hay, B. A. (2011). *MEDEA* Selfish Genetic Elements as Tools for Altering Traits of Wild Populations: A Theoretical Analysis. *Evolution*. 65, 1149–1162. https://doi.org/10.1111/j.1558-5646.2010.01186.x

Watanabe, T., Noji, S. and Mito, T. (2017). Genome Editing in the Cricket, Gryllus Bimaculatus. In *Genome Editing in Animals*. New York: Humana Press, 219–233.

Wedell, N., Price, T. A. R. and Lindholm, A. K. (2019). Gene Drive: Progress and Prospects. *Proceedings of the Royal Society B: Biological Sciences*. 286(1917). https://doi.org/10.1098/rspb.2019.2709

Wicker, T., Sabot, F., Hua-Van, A. et al. (2007). A Unified Classification System for Eukaryotic Transposable Elements. *Nature Reviews Genetics*. 8, 973–982. https://doi.org/10.1038/nrg2165

Winterberg, S., Shachar, C., Lunshof, J. and Grolman, J. (2019). *Genome Editing-Technology Factsheet Series*. Cambridge, MA: CRCS Center for Research on Computation and Society Harvard.

Wojtal, D., Kemaladewi, D., wi, U. and Malam, Z. (2016). Spell Checking Nature: Versatility of CRISPR/Cas9 for Developing Treatments for Inherited Disorders. *Am Journal Human Genetics*. 98, 90–101.

Wolfs, J. M. (2017). Dual-Active Genome-Editing Reagents. *Electronic Thesis and Dissertation Repository*. 4546. https://ir.lib.uwo.ca/etd/4546

Woo, J., Kim, J., Kwon, S., Kwon, S. I., Corvalán, C., Seung, W. C., Kim, H., Kim, S., Kim, S., Choe, S. and Kim, J. (2015). DNA-Free Genome Editing in Plants with Preassembled CRISPR-Cas9 Ribonucleoproteins. *Nature Biotechnology*. 33, 1162–1164. https://doi.org/10.1038/nbt.3389.

World Trade Organization [WTO]. (2018). *International Statement on Agricultural Applications of Precision Biotechnology*. https://docs.wto.org/dol2fe/Pages/FE_Search/ExportFile.aspx?id=249267&filename=q/G/SPS/GEN1658R2.pdf [Accessed November 20, 2018].

Wright, W. D., Shah, S. S. and Heyer, W. D. (2018). Homologous Recombination and the Repair of DNA Double-Strand Breaks. *Journal of Biological Chemistry*. 293, 10524–10535.

Xiao, Y., Luo, M., Dolan, A. E., Liao, M. and Ke, A. (2018). Structure Basis for RNAguided DNA Degradation by Cascade and Cas3. *Science*. 361. https://doi.org/10.1126/science.aat0839

Xie, F., Ye, L., Chang, J. C., Beyer, A. I., Wang, J., Muench, O. and Kan, Y. W. (2014). Seamless Gene Correction of β-Thalassemia Mutations in Patient-Specific iPSCs Using CRISPR/Cas9 and piggyBac. *Genome Research*. 24, 1526–1533. https://doi.org/10.1101/gr.173427.114

Xu, S., Pham, T. P. and Neupane, S. (2020). Delivery Methods for CRISPR/Cas9 Gene Editing in Crustaceans. *Marine Life Science and Technology*. 2(1), 1–5. https://doi.org/10.1007/s42995-019-00011-4

Yao, Y. (2018). Genome Editing: From Tools to Biological Insights. *Genome Biology*. 19(1).

Yarrington, R. M., Verma, S., Schwartz, S., Trautman, J. K. and Carroll, D. (2018). Inhibit Target Cleavage by CRISPR-Cas9 in Vivo. *Proceedings of the National Academy of Sciences of the United States of America*. 115, 9351–9358. https://doi.org/10.1073/pnas.18100 62115.

Yusorf, W. (2018). CRISPR/CAS9: An Introduction to Genome Editing. *Malaysian Journal of Paediatrics and Child Health*. 23(1).

Zafer, M., Horvath, H., Mmeje, O., van der Poel, S., Semprini, A. E. and Rutherford, G. (2016). Effectiveness of Semen Washing to Prevent Human Immuno Deficiency Virus (HIV) Transmission and Assist Pregnancy in HIV-Discordant Couples: A Systematic Review and Meta-Analysis. *Fertile Sterile*. 105(3), 645–655. https://doi.org/10.1016/j.fertnstert.2015.11.028

Zahur, M., Tolö, J., Bähr, M. and Kügler, S. (2017). *Long-Term Assessment of AAV-Mediated Zinc Finger Nuclease Expression in the Mouse Brain*, 1–13. https://doi.org/10.3389/fnmol.2017.00142

Zhang, D., Lees, R. S., Xi, Z., Bourtzis, K. and Gilles, J. R. L. (2016). Combining the Sterile Insect Technique with the Incompatible Insect Technique: III-Robust Mating Competitiveness of Irradiated Triple *Wolbachia*-infected *Aedes albopictus* Males under Semi-Field Conditions. *PLoS One*. 11(3), 1–15. https://doi.org/10.1371/journal.pone.0151864

Zhang, Y., Heidrich, N., Ampattu, B. J., Gunderson, C. W., Seifert, H. S., Schoen, C., Vogel, J. and Sontheimer, E. J. (2013). Processing-Independent CRISPR RNAs Limit Natural Transformation in *Neisseria meningitidis*. *Molecular Cell*. 50(4), 488–503. https://doi.org/10.1016/j.molcel.2013.05.001

Zhang, Y., Massel, K. and Godwin, I. D. (2018). Applications and Potential of Genome Editing in Crop Improvement. *Genome Biology*. 19, 210.

Zhang, Z., Zhang, Q. and Ding, X. D. (2011). *Advances in Genomic Selection in Domestic Animals.* https://link.springer.com/article/10.1007/s11434-011-4632-7

Zheng, X., Zhang, D., Li, Y., Yang, C., Wu, Y., Liang, X., Xi, Z. et al. (2019). Incompatible and Sterile Insect Techniques Combined Eliminate Mosquitoes. *Nature*. 572(7767), 56–61. https://doi.org/10.1038/s41586-019-1407-9

Zhou, W. and Deiters, A. (2016). *Conditional Control of CRISPR/Cas9 Function*. https://pubmed.ncbi.nlm.nih.gov/26996256/

Zuo, E. (2019). Cytosine Base Editor Generates Substantial Off-Target Single-Nucleotide Variants in Mouse Embryos. *Science*. 364, 289–292.

Zuyong, H., Chris, P., Bruce, A. and Simon, G. L. (2016). *Comparison of CRISPR/Cas9 and TALENs on Editing an Integrated EGFP Gene in the Genome of HEK293FT Cells.* https://springerplus.springeropen.com/articles/10.1186/s40064-016-2536-3

Index